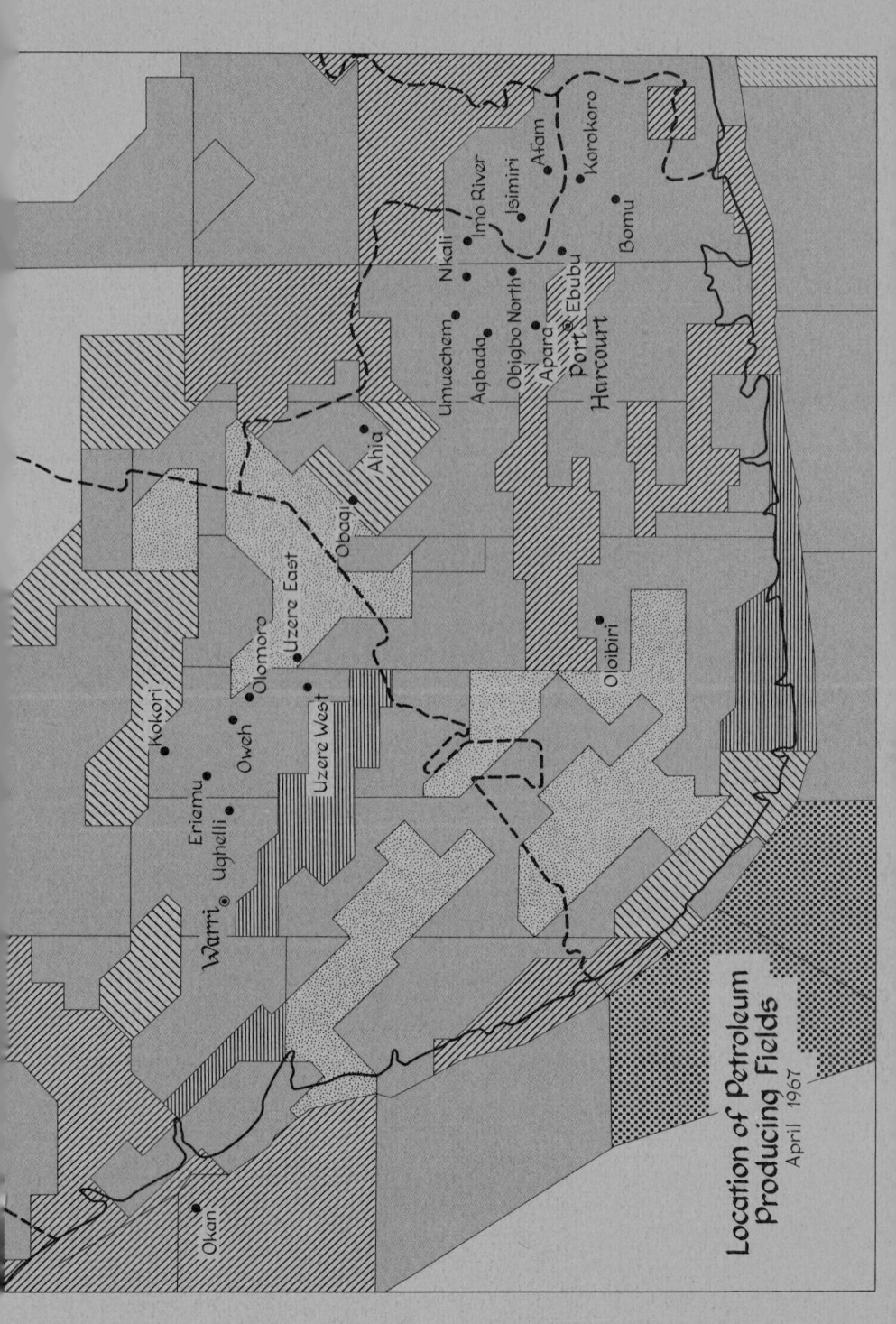

Petroleum and the Nigerian Economy

Petroleum and the Nigerian Economy

Scott R. Pearson

Stanford University Press, Stanford, California 1970

Stanford University Press
Stanford, California
© *1970 by the Board of Trustees of the*
Leland Stanford Junior University
Printed in the United States of America
ISBN 0-8047-0749-9
LC 76-130830

To Sandra

Preface

A prerequisite of contemporary empirical research is the generous cooperation of individuals in the particular institutions under study. My greatest debt is thus owed to countless Nigerian government officials and representatives of the petroleum companies active in Nigeria. Few, if any, of these persons can be expected to endorse all of the conclusions reached in this study. Most, therefore, desire to remain anonymous. I am deeply grateful for their invaluable assistance.

My learning about petroleum and Nigeria has been supported and enhanced by affiliation with several organizations. My initial direct contact with Nigeria took place in 1961–63, when I was a member of the first group of Peace Corps Volunteers in that country. I made several shorter visits to Nigeria in 1965, 1966, 1968, and 1969 as an economic consultant to the U.S. Agency for International Development's Mission to Nigeria. During the first two of these consultancies, I had the pleasure of collaborating with Wilson E. Schmidt, who has continuously encouraged and assisted my analysis of Nigerian petroleum. In 1967–68, I wrote an earlier version of this study as my doctoral dissertation under the direction of Albert O. Hirschman and Walter P. Falcon; I am grateful to both for their instructive guidance. In addition, I wish to thank Resources for the Future and the Ford Foundation for dissertation support. Finally, I must acknowledge the excellent professional and institutional support that I have received from my colleagues and the staff of the Food Research Institute during the past two years. I am especially indebted to William O. Jones, the Director

of the Institute, and to Bruce F. Johnston, Benton F. Massell, and C. Peter Timmer for their thoughtful criticism and constant encouragement. Naturally none of these institutions and individuals necessarily agrees with my analysis or conclusions. More specifically, this book does not represent the official or unofficial opinions of any U.S. government agency.

In addition to those persons already mentioned, several other social scientists have kindly assisted my research or given me comments on earlier drafts of this book. A partial listing of this group includes Michael H. B. Adler, Robert O. Blucker, Samuel Bowles, Anson Chong, John B. Cownie, Thomas Gewecke, Gerald K. Helleiner, Donald B. Keesing, Peter Kilby, Hayne E. Leland, Alan S. Manne, Stanley A. Nicholson, W. Haven North, Charles S. Pearson, Ralph Pochoda, Thomas Rawski, Sherman Robinson, Ludwig H. Schatzl, John Shilling, Alan Sokolski, and Thomas R. Stauffer.

Many others had a direct hand in my efforts. My research assistant, John M. Page, Jr., was equally proficient at programming the computer, chasing down minute details, and offering constructive suggestions; I am particularly grateful for his long and patient service. My secretary, Eva Langemyr Larsen, cheerfully engineered the various drafts of my manuscript to completion with unusual efficiency and meticulousness. Rosamond Peirce greatly improved the content and readability of the numerous tables in the book, Sue Riggs painstakingly constructed the index, and Jill Leland contributed the excellent cartography. And perhaps most important, my family—Sandra, Sarah, and Elizabeth—has endured my absences of mind and body that were caused at least in part by my desire to write this book.

S.R.P.

September 25, 1970

Contents

	Explanation of Measures and Symbols	xv
	Introduction	1
ONE	The International Petroleum Industry	5
TWO	Petroleum Company Operations in Nigeria	12
THREE	Recent Trends in Nigerian Economic Development	31
FOUR	Foreign Investment in a Less Developed Economy	39
FIVE	The Direct Contribution of Petroleum to Income	55
SIX	Factor Contributions to Other Sectors	70
SEVEN	Linkage Effects	86
EIGHT	Petroleum in Nigeria: Some Projections	104
NINE	The Politics of Nigerian Oil	137
TEN	Implications for Policy	153

Appendixes

A. National Income and Balance of Payments, 169.
B. Payments of Companies to the Government, 176.
C. Subsidiary Firms of the Nigerian Petroleum Industry, 189. D. The Input-Output Data, 195. E. The Petroleum Industry's Value Added and Balance-of-Payments Impact, 205.

Notes	209
Selected Bibliography	221
Index	229

Tables

1.1	Petroleum Production and Exports: World Totals, Selected Years 1918–68	6
1.2	World Demand for Energy, 1960 and 1980	6
1.3	Petroleum Production and Payments to Government, Selected Countries, 1950 and 1965	10
2.1	Petroleum Exploring and Producing Companies Holding Nigerian Concessions, April 1969	16
2.2	Exports of Crude Petroleum from Nigeria, 1966	19
3.1	Sectoral Contribution to GDP (at Factor Cost), Selected Years 1950–68	32
3.2	Composition of Domestic Exports, Selected Years 1900–1968	34
3.3	F.O.B. Export Prices and Producer Prices for Major Nigerian Agricultural Exports, 1957–63	37
3.4	Ratios of Producer Price to Food Price	37
5.1	Crude Petroleum Production and Exports, 1958–68	56
5.2	Petroleum Industry Value Added, 1963–68	57
5.3	The Contribution of Petroleum Value Added to Nigerian GDP and to Nigerian GNP, 1963–68	58
5.4	Leontief Production Function for the Petroleum Industry, 1965 Data	60
5.5	Comparison of Factor Shares for the Petroleum Industry, 1963–68	61

5.6	Input Coefficients, 1965 Data	62
5.7	Leontief Production Function for the Consolidated Nigerian Petroleum Industry, 1965 Data	63
5.8	Comparison of Extractive Export Industry Input Coefficients for Selected Industries and Countries	64
6.1	Petroleum Industry Payments to the Nigerian Government, 1959–62	72
6.2	Petroleum Industry Payments to the Nigerian Government, 1963–68	72
6.3	Comparison of Petroleum and Nonpetroleum Nigerian Government Revenues, 1959–68	73
6.4	Nigerian Governments' Public Saving, 1959–66	74
6.5	Impact of the Petroleum Industry Balance of Payments, 1963–68	76
6.6	Foreign Exchange Availability and Use, 1963–68	78
6.7	Total Imports Associated with Operations of the Nigerian Petroleum Industry, 1965–68	81
6.8	Import Coefficients for Industries Supplying or Servicing the Petroleum Industry, 1965 Data	82
6.9	Levels of Employment in the Nigerian Petroleum Exploring and Producing Industry, March 31, 1964–67	84
7.1	Selected Ancillary Firms' Receipts from the Petroleum Industry Paid Abroad and Paid Locally by Ownership of Firms, 1965 Data	88
7.2	Selected Ancillary Firms' Receipts Paid Locally from All Industries and the Share from the Petroleum Industry, 1965 Data	90
7.3	Petroleum Refinery Operations, January–April 1967	93
7.4	Industries and Utilities Using Natural Gas as a Source of Energy, 1967	96
7.5	Total Payments to Government Associated with the Production of Nigerian Petroleum, 1965	102

Tables xiii

8.1	Petroleum Production, 1963–73	106
8.2	Other Local Payments, 1963–73	108
8.3	Petroleum Industry Payments to Government, by Category, 1963–73	109
8.4	The Impact of Petroleum Revenues on Total Government Revenues, 1963–73	110
8.5	Balance-of-Payments Impact of the Petroleum Industry, 1963–73: Local Currency Expenditures	112
8.6	Balance-of-Payments Impact of the Petroleum Industry, 1963–73: International Financial Flows	113
8.7	Balance-of-Payments Impact of the Petroleum Industry, as Per Cent of Petroleum Export Earnings, 1963–73	114
8.8	Petroleum Industry Contribution to Nigerian Foreign Exchange Availability and Use, 1963–73	115
8.9	Value Added of the Petroleum Industry, 1963–73	117
8.10	Contribution of Petroleum Value Added to Nigerian Gross Domestic Product, 1963–73	120
8.11	Contribution of Petroleum Value Added to Nigerian Gross National Product, 1963–73	121
8.12	Input-Output Model: Summary of Results Using High Projection	126
8.13	Input-Output Model: Summary of Results Using Low Projection	130
9.1	Crude Petroleum Production by Field and State, April 1967	140
9.2	Ethnic Origin and State Location of Nigerian Petroleum Production, April 1967	141
9.3	Hypothetical Example of Origin and Distribution of Actual 1967 Petroleum Industry Payments to the Nigerian Government	144
9.4	The Division of Petroleum Production Between Biafra and Mid-Western Nigeria, 1969–73	147

9.5	The Impact of Petroleum Revenues on Biafran Government Revenues, 1969–73	147
9.6	Biafran Petroleum Industry Balance-of-Payments Impact, 1969–73	148
9.7	Petroleum Industry Contribution to Biafran Foreign Exchange Availability and Use, 1969–73	149
9.8	The Contribution of Petroleum Value Added to Biafran Gross National Product, 1969–73	150

Explanation of Measures and Symbols

The following equivalents may be helpful:
 1 £ Nigerian = U.S. $2.80, 1950–69
 1 long ton = 2,240 pounds
 1 metric ton = 2,204.6 pounds
 1 petroleum barrel = 42 U.S. gallons
 1,000 liters = 6.2898 42-gallon barrels

Since the specific gravity of crude petroleum varies from country to country and from year to year, conversions between weight and capacity cannot be accurate. The United Nations *Statistical Yearbook, 1966*, p. 202, uses a specific gravity of .85 for both the United States and Nigeria, resulting in the following conversion factors:

 1 barrel = approximately .135 metric tons
 1 metric ton = approximately 7.4 barrels
 1 metric ton per year = approximately .020 barrels per day

The following symbols are used in the tables:
 ... Data not available, or category not applicable
 — Magnitude zero, or less than one-half the unit employed

Because of rounding, totals may not exactly equal sum of items shown.

Petroleum and the Nigerian Economy

Introduction

Whenever a large private foreign investor undertakes significant economic activity in a less developed country, there are bound to be far-reaching effects. Whether or not these effects result in a net gain over time to the society concerned is a question of considerable interest and one that has been the topic of intense debate for several decades. The purpose of this study is to throw additional light on this issue in two principal ways. The first involves construction of a general analytical framework to evaluate economic benefits and costs that might be associated with private direct foreign investment in a low-income economy. The second concerns the application of this framework to the recent operations of international petroleum corporations in Nigeria.

In the decade centered on the attainment of independence in 1960, Nigeria appeared to be progressing admirably. Nigeria's rate of economic development, as measured in the aggregate by growth of income, was respectable and steady. But beneath this deceivingly prosperous surface, the Nigerian political system was threatened by a series of crises in which the principal antagonists repeatedly went to the brink yet always managed to compromise short of disaster. During this decade, changes occurred that had wide economic and political ramifications. The discovery (in 1956) and production (from 1958) of crude petroleum were not immediately some of the more important of these. Economic benefits associated with Nigerian oil only began to appear near

the end of this period and then with limited significance. As the economic importance of petroleum received wider exposure, it naturally became an increasingly crucial political issue.

In 1966 the seemingly stable situation began to disintegrate visibly. The political system crumbled under the weight of two military coups. But the resilient Nigerian economy continued to grow although with somewhat less buoyancy. Biafran secession and the resultant civil war in mid-1967 completed the destruction of the prior political structure and severely jolted certain parts of the economy. The tragic and destructive civil war lasted for two and one-half years before the federal government emerged victorious in January 1970. Following the end of the fighting, the long, slow process of reconstruction and rehabilitation began.

By the mid-1960's petroleum was a significant and growing social force. For the first time in its history in less developed areas, the production of petroleum had been superimposed on a diversified and growing economy, but one supported at that time by a political structure of demonstrably questionable viability. What are the recent and likely future impacts on the Nigerian economy of the flow of petroleum-related investment into Nigeria? This is the central question to which this book is addressed.

An answer to this question requires construction of a framework of analysis that will facilitate a thorough understanding of what has already taken place and of a model that will enhance the reliability of speculations about the future. The methodology developed in Chapter Four to analyze the effects of petroleum in Nigeria is in general applicable to all instances in which powerful private foreign investors initiate structural changes in less developed economies. In this regard the first of the two principal objectives of this study is fulfilled.

Achievement of the second objective demands significantly more space and energy. To be relevant and interesting, empirical economic analysis must contain a balanced intermixture of institutional details and theoretical models. Analysis of the effects of investment in Nigerian petroleum thus begins with a detailed de-

Introduction

scription of the appropriate institutional setting. The first three chapters of this study are devoted to this topic. As a first step, in Chapter One the international institutions of the Western petroleum corporations, especially certain peculiarities that have a major influence on less developed petroleum-producing economies, are described briefly. In Chapter Two the evolution of institutional arrangements surrounding the production of petroleum in Nigeria is summarized. Major attention is given to factors that influence the decision-making processes of the Nigerian government, on one hand, and of the petroleum companies, on the other. The laying of the institutional groundwork is completed in Chapter Three with a very brief review of the most important recent developments in the Nigerian economy.

In the second part the reader is provided with a sharp break. Neither petroleum nor Nigeria is mentioned explicitly in Chapter Four. Instead the theoretical approach that is followed in succeeding portions of the book is set out. This methodological apparatus includes a general analytical framework to gauge the economic effects of direct private foreign investment on the economy and an input-output model to estimate the possible future importance of the major domestic effects of the capital inflow.

With the institutional ground work and theoretical approach in hand, the empirical analysis begins in Chapter Five where the net direct contribution to Nigerian domestic product of investment in petroleum production is estimated. To accomplish this, a Leontief production function is calculated and assumptions about the opportunity costs of factors employed are outlined explicitly. Chapters Six and Seven are concerned with indirect benefits and costs associated with foreign investment in Nigerian petroleum. Chapter Six treats the petroleum industry's contributions of factors of production to the non-oil economy, and Chapter Seven considers the impacts of intersectoral relationships on total Nigerian economic progress. In light of the rapidly changing conditions surrounding Nigerian petroleum, considerable interest attaches to Chapter Eight, where projections of likely future effects of oil in

Nigeria are undertaken within the framework of an input-output requirements analysis.

In the last section the focus shifts to policy considerations. Some political implications of petroleum in Nigeria are set forth in Chapter Nine. That discussion serves as a prelude to the final chapter, Chapter Ten, which contains a brief integration and summary and some implications for policy.

CHAPTER ONE

The International Petroleum Industry

Crude Petroleum Production and Exports

Before private foreign investment will enter a less developed country, evidence must exist of a large and growing external demand sufficient to absorb incremental production by the country under examination.[1] The demand for crude petroleum is very large and has been growing rapidly during the past decade—at an annual rate of 7½ per cent from 1958 to 1968.[2] Table 1.1 displays world totals for crude oil production and exports and demonstrates that exports are steadily assuming an increasingly large portion of total production. Over 90 per cent of traded crude is exported from low income countries.[3]

Barring major technological shifts and unforeseen governmental policy changes, the long-standing, buoyant external demand for crude oil should continue. An examination of the possible prospects for a continuation of this attractive picture naturally requires treatment of the total demand for energy as well as the role that petroleum might play within this global picture. Accordingly, Table 1.2 is based on a recent projection by the Organisation for Economic Co-operation and Development of world energy demand. Although the total demand for energy is projected to increase more than two and one-half times between 1960 and 1980, that for petroleum products is expected to rise more than three times. Reference to Table 1.1 shows that petroleum production (and consumption) rose by 83 per cent between 1960 and 1968. Petroleum output might be expected to be at least two-thirds again as large as the 1968 level by 1980. Though energy projections are

TABLE 1.1. PETROLEUM PRODUCTION AND EXPORTS:
WORLD TOTALS, SELECTED YEARS 1918–68
(*Million barrels per day*)

Year	Production	Exports	Exports as percentage of production	Year	Production	Exports	Exports as percentage of production
1918	1.4	...[a]	...	1962	25.5	11.0	43
1928	3.6	...[a]	...	1963	27.4	12.2	45
1938	5.4	.9	17	1964	29.5	13.6	46
1948	9.4	2.4	26	1965	31.7	15.0	47
1958	18.9	7.3	39	1966	34.4	16.6	48
1959	20.5	7.9	39	1967	37.0	17.9	49
1960	22.0	8.8	40	1968	40.2	20.4	51
1961	23.5	9.8	42				

Data from *Statistical Review of the World Oil Industry, 1968* (The British Petroleum Company Limited, London, 1969), pp. 18–21; and Gabriel, p. 41. (Frequently cited sources in the tables are short forms. Complete titles, authors' names, and publication data will be found in the Selected Bibliography, pp. 221–26.) Figures shown include data for the USSR and Eastern Europe and estimates for China.

[a] Comparable data are not available.

TABLE 1.2. WORLD DEMAND FOR ENERGY, 1960 AND 1980
(*Million metric tons petroleum equivalent*[a])

	1960			1980		
Demand	OECD	Non-OECD	World total	OECD	Non-OECD	World total
Petroleum	738	330	1,068	2,150	1,300[b]	3,450
Nonpetroleum	1,105	932	2,037	2,150	2,600	4,750
Total energy demand	1,843	1,262	3,105	4,300	3,900	8,200

Data from *Energy Policy*, pp. 17–42, 98–99, 150–51. OECD countries are Austria, Belgium, Canada, Denmark, France, Federal Republic of Germany, Greece, Iceland, Ireland, Italy, Japan, Luxembourg, the Netherlands, Norway, Portugal, Sweden, Spain, Switzerland, Turkey, the United Kingdom, and the United States.

[a] For rough conversion to thousand barrels per day, multiply by 20.

[b] Author's approximation, at 33⅓ per cent of the estimate of total energy demand.

notoriously hazardous, there seems to be little room for concern about inadequate demand for petroleum in the foreseeable future. The OECD study concludes that sufficient oil reserves are being developed to meet this demand.[4]

In moving petroleum from the ground to the consumer, the oil industry performs four separate but vertically integrated functions—production, transportation, refining, and marketing. In the main, the industry concentrates its activity under each of the last three headings in the economically more advanced nations. These

downstream functions are thus of only passing concern in this discussion. Production, the activity of principal interest to those underdeveloped countries fortunate enough to have petroleum deposits, is conveniently subdivided into four chronological phases—pre-drilling activities, exploratory drilling, development drilling and equipping of wells, and production itself.[5]

The nature of the functions performed by the petroleum industry has had a very definite impact on its industrial organization. In general, the industry has been characterized by bigness and backward vertical integration from the user to the source.[6] This has resulted from a combination of the following factors: the tremendously high initial capital requirements; the implied need to coordinate precisely so as to reduce nonutilization of capacity; the large exploration risks and variable timing factors that require multi-area operation; and the desire to have flexibility to shift in or out of one area as it becomes more or less economically or politically attractive relative to other areas.

Structure of Crude Transfer Prices

The current disequilibrium cost-price structure in the international oil industry was originally established by the major corporations as a rational means of serving their own goals. Ironically, over time changing circumstances have locked the industry into what has become, in its view, an increasingly onerous system because of the incentive it inadvertently created for host country governments.[7] As a result there are wide margins between the industry's production costs and the prices it must use to calculate payments to governments. These margins reflect the economic rent involved in oil production and hence create a large host country tax base.* The oil industry thus pays significantly greater amounts

---

* In the vertically integrated international petroleum industry, each company decides the prices at which to evaluate crude production from each of its sources of supply. These evaluations are known as crude transfer prices. Originally these prices posted by each company served as the basis for tax obligation calculations. But during the past decade nearly all less developed producing countries have negotiated posted prices with the industry for tax calculation purposes so as to preclude any reduction of per barrel tax revenues due to softening of crude transfer prices.

of taxes to host governments of producing countries than would be the case if prices were more closely reflective of production costs and if taxes were based on a normal concept of profit.

It is of interest to examine briefly the industry's original predilection in favor of high crude transfer prices and how underlying circumstances have altered to erode earlier justifications for high prices.[8] Two decades ago the oil industry was motivated to establish and maintain a system of high crude transfer prices for the following principal reasons. First, prior to the emergence of substantial production in less developed areas, high-cost U.S. crude served as the main source of internationally traded oil. Because exports from the new, less developed areas of production took place mainly within established vertically integrated corporations, there was no incentive to lower prices below the U.S. export price plus transport differential, so long as the United States was still a net crude exporter. Second, trade and payments regulations in the importing countries existed to protect domestic energy suppliers and conserve scarce foreign exchange. If the companies had set lower crude export prices, this would only have resulted in increased protection on the part of importing nations. In addition, the widespread postwar currency regulations created an incentive for the industry to generate large flows of convertible currencies for reinvestment purposes. Third, there were taxation advantages in maintaining high crude transfer prices. On one hand, until 1949 the incremental tax rate for the vertically integrated firm was generally lower in the producing countries than in the consuming areas. On the other hand, high crude prices raised the base on which U.S. companies could calculate percentage oil depletion allowances.[9] And finally, the companies found it useful to set and maintain high crude transfer prices in order to retain the oligopolistic nature of the industry. For technical reasons already mentioned, marketing was the easiest function of the industry for outsiders to enter. By transferring accounting profits from downstream operations to the producing end, the vertically integrated companies put pressure on outsiders wishing to crack the oligopoly at its most vulnerable point.

The International Petroleum Industry

Initially the host country governments were relatively uninformed and unsophisticated in their dealings with the international oil industry. But it did not take them long to recognize the scope for gain that was inherent in the disequilibrium cost-price situation. Starting in the late 1950's, circumstances encroached upon and eventually overtook oil company rationale for the high crude transfer price structure and at the same time severely circumscribed the industry's ability to shift policy. In the first place, many of the most restrictive trade and payments regulations in crude petroleum importing countries were decreased or eliminated. Second, the host country governments with industry agreement instituted a 50 per cent income tax rate and then gradually increased the effective rate of taxation on petroleum production. In large part this has merely shifted the tax incidence from each oil company's home country to the producing country in question because of the existence of home country tax credits. Moreover, producing country governments placed a floor under the high price structure by instituting arrangements calling for the calculation of taxes on the basis of negotiated posted prices rather than company crude transfer prices. Third, the international majors have been confronted with an increase in international competition in all functional stages, especially production. As evidence of this, between 1950 and 1965 the majors' share of production (excluding the United States and the USSR) dropped from nearly 90 per cent to 75 per cent, while the number of companies with at least 1,000 barrels per day production increased from 57 in 1957 to 104 in 1965.[10] For these reasons the cost-price structure evolved by the petroleum companies some two decades ago has in recent years increasingly benefited host country governments as well as the companies themselves at the expense of consumers in importing countries.

Impact on Producing Country Growth Prospects

Interest in the institutional structure of the international petroleum industry derives from its possible influence on the industry's impact on a producing country's economy. Table 1.3 documents

TABLE 1.3. PETROLEUM PRODUCTION AND PAYMENTS TO GOVERNMENT,
SELECTED COUNTRIES, 1950 AND 1965

Country	Production (thousand barrels per day) 1950	1965	Payments to government Total (million U.S. dollars) 1950	1965	Per barrel (U.S. cents) 1950	1965	Annual percentage growth rates, 1950-65 Production	Total payments to government
Venezuela	1,492	3,473	330	1,128	61	89	5.8	8.5
Saudi Arabia	547	2,025	113	639	57	87	9.1	12.2
Kuwait	344	2,170	12	636	10	80	13.1	31.0
Iran	661	1,886	91	532	38	77	7.2	12.5
Iraq	128	1,315	19	368	41	77	16.8	21.9
Total, five countries	3,172	10,869	565	3,303	49	84	8.6	12.5

Data from *World Oil*, August 15, 1951 and 1966; and from *Petroleum Press Service*, September, 1962 and 1966; both cited in Gabriel, p. 65.

the results of the process whereby rapidly growing petroleum exports have combined with financial arrangements shifting in favor of the producer governments to cause large and increasing financial payments to the producing countries. The fact that these payments invariably go to host country governments has resulted in a shift of economic power from the private toward the public sector in most oil-producing countries. It is ironic that the mainly private Western corporations greatly enhance the power of recipient governments by making substantial payments that are often used to enlarge the local public sector absolutely as well as relative to the private sector.

The impact of the institutions surrounding petroleum exports on the producing country's private sector is less certain. The international oil companies and the producing country governments and private factor owners have undergone a mutual learning process. Industry technology is only slightly though continually altered by host country pressures to attain goals at variance with those of the industry. One international development that did have large consequences for private sectors in host countries was the gradual elimination of the barter policy that emerged after World War II, whereby certain advanced nation importers of crude petroleum insisted on swapping oil industry machinery and spare parts

for crude to meet their own balance-of-payments exigencies. The aggregate effects of the oil industry's organization on private sectors of host countries are very difficult to generalize about and are thus probably best investigated on a case-by-case basis.

As a means of summarizing the main results of this section, it is illuminating to review the major tenets of a significant group of economists who are in general pessimistic concerning the economic benefits of private foreign investment in extractive export industries.[11] One can then determine whether their ideas are applicable to petroleum. The pessimists claim, first, that international demand for primary exports is no longer increasing. Second, they allege that powerful monopoly mechanisms favor the more advanced countries in those export markets that are firm. And third, so the argument runs, it would make little difference if exports could grow rapidly, since contemporary conditions seriously impede the transmission of growth from the extractive export sector to the remainder of the low income economy.

Contrary to the fears of these pessimists, external demand for petroleum has been increasing steadily and is expected to continue to do so in the foreseeable future. As for the problem of monopoly elements, the net result of the interaction between the oligopolistic controls employed by the oil companies on one side and the producing governments on the other has been a larger stream of payments to producing governments than is likely to have occurred in the absence of oligopoly.[12] Finally, regarding the pessimists' claim that foreign-owned extractive export industries in particular are unable to transmit growth to other sectors of host underdeveloped economies, the process of growth via the contribution of factors and linkage effects is best investigated on a country-by-country basis before generalizations can be made.[13] This study of Nigerian petroleum can be viewed as one effort toward that end.

CHAPTER TWO

Petroleum Company Operations in Nigeria

The exploration for and the production of crude petroleum are many-faceted operations. It would be highly unusual for any single analyst to possess the wide-ranging capabilities that might be required to probe and evaluate the physical as well as the social ramifications of this process. The scope of this study comprises economic as well as associated political aspects of the operations and impacts of the petroleum industry in Nigeria. Physical characteristics of Nigerian petroleum production are included only when this adds economic insight.[1]

This chapter sets the institutional stage for a detailed analysis of economic impacts by discussing the operations of petroleum companies in Nigeria and summarizing their financial arrangements with the Nigerian government. There are three related institutional themes—each of which concerns company decision-making—that permeate this chapter and indeed the entire study. The first is the uncertainty regarding production levels that emanates from intra-company decision-making. The head office of each international petroleum company active in Nigeria sets and resets production levels for its Nigerian subsidiary on the basis of highly complicated criteria. The second is the by-play between and among the operating subsidiaries in Nigeria. The alternation between cooperation and harsh acrimony among oil companies is certainly not peculiar to the Nigerian petroleum scene, but it does cause complications in what is often alleged to be an imperturbably stable industry. The evolving interrelationship between the industry on one hand and the Nigerian government on the other, the third

theme, is surely the most important. This mutual learning process continues, with results that are nearly always significant. The operation of each petroleum production company in Nigeria is thus constrained and heavily influenced by its dealings with three institutions—its head office, the remainder of the industry in Nigeria, and the Nigerian government.

The Petroleum Companies in Nigeria

There are basically four methods by which a technically less advanced country can tap a potential extractive export reserve: (1) invite foreign concessionaires to form local subsidiaries and thereby supply management and technology, capital, and markets; (2) undertake joint ventures in which foreign investors supply management and technology and markets as well as a portion of the capital—with the remaining capital furnished by the host country; (3) institute management contracts whereby experienced foreign firms offer management and technology only and the host supplies the capital and tries to find markets; or (4) do without foreign participation altogether and furnish management and technology, capital, and markets itself.[2] The British colonial government of Nigeria, as sole owner of all subsoil minerals in Nigeria, chose and implemented the first of these alternatives.* Moreover, the policy of inviting foreign petroleum companies to search for and produce oil in Nigeria was continued by independent Nigerian governments that decided to limit this participation for the most part to large Western corporations. In view of the limited Nigerian technical expertise, the difficult technology and the large capital requirements associated with oil production, the alternative productive opportunities available for Nigerian investment resources,

* It could be argued that there was little real choice in view of the seemingly poor prospects of discovering petroleum in Nigeria before the mid-1950's. For example, L. Dudley Stamp, the noted British geographer, wrote in 1953: "Apart from the fields along the shores of the Gulf of Suez in Egypt, and a small yield from three tiny fields in Algeria and four in Morocco, Africa has no oil. The conditions favoring the accumulation of oil in quantity, in folds among sedimentary rocks on the margins of great sedimentary basins, do not exist in Africa." See L. Dudley Stamp, *Africa: A Study in Tropical Development* (New York: Wiley, 1953), p. 53.

and the restricted nature of the market for nonintegrated crude, these decisions are easily explained. But this is not to imply that they are necessarily optimal, as later discussion makes clear.

All of the seven international majors are now or have recently been active in Nigeria. Of this group only Standard of New Jersey does not currently have a Nigerian concession. The international minors in Nigeria include Italian and French government-owned concerns as well as private American corporations. Of particular interest is the sharing agreement between the Nigerian Agip Oil Company (owned equally by Agip, an Italian government corporation, and Phillips, an American private corporation) on the one hand and the Nigerian government on the other. This agreement allows the federal government to purchase up to 30 per cent of the company's share capital in installments to be paid over three years and to pass on its shares to state governments or public corporations. To date the government has chosen not to exercise its option, though there have been indications that it will certainly do so once the profitability of the Nigerian Agip operations is fully guaranteed. It is significant that in the Petroleum Decree 1969, the provisions of which apply only to future concessions, the possibility of participation by the federal military government on terms to be negotiated is specifically allowed for if it is deemed to be in the public interest (see Appendix B).

Table 2.1 draws together assorted pieces of information pertaining to the international petroleum companies that are currently exploring for crude oil and in four instances producing in Nigeria.* Nine companies hold Nigerian concessions amounting to a total of 44,857 square miles, and two more have applied for acreage. The location of these concessions is shown on the end sheets. On this map the blocks off the southern coast of Nigeria locate offshore concessions.

The formation of this concession pattern is worth relating. The

---

* Monthly production in April 1970 exceeded 1,000,000 barrels per day (b/d). Companies producing at that time were Shell-BP, Gulf, Texas Overseas, and Mobil. Safrap, the only other company that had gone into production in Nigeria before then, was forced to stop producing when the civil war began in July 1967.

search for oil in Nigeria began in 1908 when a German company, Nigerian Bitumen Corporation, drilled fourteen wells in what is today Lagos State before ceasing operations with the outbreak of World War I.[3] Interest in the possibility of discovering oil in Nigeria revived in 1937 with the establishment of Shell/D'Arcy Exploration Parties, a consortium owned equally by Royal Dutch Shell and British Petroleum that later became the Shell-BP Petroleum Development Company of Nigeria Ltd.[4] In November 1938 this company received an Oil Exploration License (OEL) covering all of Nigeria. By 1957 Shell-BP had reduced its acreage to 40,000 square miles of Oil Prospecting Licenses (OPL's). Of this acreage Shell-BP converted nearly 15,000 square miles into Oil Mining Leases (OML's) in 1960 and 1962 and returned the residual to the Nigerian government.[5]

Between 1938 and 1941 Shell-BP undertook preliminary geological reconnaissance. After a five-year interruption caused by World War II, it intensified and followed up this activity with geophysical surveys in the 1946–51 period. In 1951 Shell-BP drilled its first wildcat well; it came up dry. During the next four years the company concentrated efforts in the Cretaceous areas rimming the Niger Delta without discovering any oil-producing wells.[6] After shifting focus to the Tertiary area of the delta itself, Shell-PB made Nigeria's first commercial discovery in 1956 at Oloibiri in what is now Rivers State. This touched off a period of extensive exploration activity in the Tertiary, which is still continuing.* Nigeria was thus ushered onto the international oil stage.

Even before Shell-BP started exporting oil from Port Harcourt in 1958, other companies began to show interest in Nigeria. Mobil carried out reconnaissance work in the northwestern corner of the country in the mid-1950's and then shifted to the coastal area in what is now Lagos State between 1958 and 1961, drilling four dry

---

* Interest and activity have shifted back to the Cretaceous area as well. In 1967 Safrap made two commercial discoveries in the Cretaceous, an oil well in the Anambra River basin and a gas well farther east at Ihandiagu. Both are on the East Central State side of the border with Kwara State. Because of interruptions caused by the civil war, the extent of deposits in this area is still a large question mark.

TABLE 2.1. PETROLEUM EXPLORING AND PRODUCING COMPANIES HOLDING NIGERIAN CONCESSIONS, APRIL 1969

Name of company	Country of original incorporation[a]	Ownership	Area (square miles)			Concessions: Oil Prospecting Licenses (OPL's) and Oil Mining Leases (OML's)[b] Location
			Onshore	Offshore	Total	
Gulf Oil Co. (Nigeria) Ltd.	U.S.A.	Gulf Oil Corp. (U.S. private), 100%	3,965	2,890	6,855	East Central State (onshore); Lagos State (offshore); Mid-Western State (offshore and onshore); Rivers State (onshore); South Eastern State (onshore); Western State (offshore)
Mobil Producing Nigeria Ltd.	U.S.A.	Mobil Oil Co. (U.S. private), 100%	—	2,025	2,025	South Eastern State
Nigeria Agip Oil Co. Ltd.	Nigeria	E.N.I. (Ente Nazionale Idrocarbum) (Italian govt.), 50% Phillips Petroleum Co. (U.S. private), 50%	2,031	—	2,031	East Central State; Mid-Western State; Rivers State
Phillips Oil Co. (Nigeria) Ltd.	U.S.A.	Phillips Petroleum Co. (U.S. private), 100%	1,401	—	1,401	Mid-Western State; Western State
Safrap (Nigeria) Ltd.	Nigeria	Société Africaine d'Exploration Pétrolière (SAFREX) (French govt.), 50% Enterprise de Recherches et d'Activités Pétrolières (ERAP) (French private), 40% Société de Gestion des Participations de la Régie Autonome des Pétroles (SOGERAP) (French private), 10%	9,336	—	9,336	Benue-Plateau State; East Central State; Kwara State; Mid-Western State; Rivers State

Company	Country	Ownership			Area (sq mi)	Location
Shell-BP Petroleum Development Co. of Nigeria Ltd.	Nigeria	Royal Dutch Shell Petroleum Co. (Dutch-British-others, private), 50%; British Petroleum Co. (British govt. and private), 50%	14,992	3,906	18,898	East Central State (onshore); Kwara State (onshore); Mid-Western State (offshore and onshore); Rivers State (offshore and onshore); South Eastern State (offshore and onshore); Western State (onshore)
Tenneco Oil Co. of Nigeria Ltd.	U.S.A.	Cumberland Corp. (wholly owned subsidiary of Tennessee Gas Transmission Co.), 50%; Sinclair International Oil Co, 25%; Sunray DX Oil Co, 25% (all U.S. private)	1,380	—	1,380	Mid-Western State; Rivers State
Texas Overseas (Nigeria) Petroleum Ltd.	U.S.A.	California Asiatic Oil Co. (U.S. private), 50%; Texas Overseas Petroleum Co. (U.S. private), 50%	—	1,931	1,931	Rivers State
Union Oil Nigeria	U.S.A.	Union Oil Co. of California (U.S. private)	—	1,000	1,000	Lagos State

Data from Nigeria [6], p. 9; updated and supplemented by unpublished individual company information. (Frequently cited Nigerian government publications are referred to by bracketed numbers and will be found in the Selected Bibliography, pp. 221–26, listed alphabetically under Nigeria.) Esso Exploration Incorporated, a subsidiary of Standard of New Jersey, which formerly held two Oil Exploration Licenses (OEL's) in what is now East Central State, South Eastern State, and Benue-Plateau State, no longer operates in Nigeria. Great Basins (Nigeria) Oil Company, a subsidiary of Great Basins Oil Company of Los Angeles, has held an OEL in what is now Benue-Plateau State, Kwara State, and North-Western State and is negotiating to convert portions of this into Oil Prospecting Licenses (OPL's). Delta Oil (Nigeria) Ltd., a Nigerian company incorporated by Godfrey Amachree, has applied for an OPL in the Mid-Western State.

[a] All exploration and production companies not previously incorporated in Nigeria were converted to Nigerian companies in 1969 to comply with Decree No. 51, Part X, which made it obligatory for all companies engaged in operations in Nigeria to be incorporated within the country.

[b] The period of Oil Mining Leases is 30 years onshore and 40 years offshore.

holes before abandoning the area.⁷ Following Shell-BP's release of acreage, Tenneco, Gulf, Agip, Safrap, and later Phillips obtained onshore OPL's. Additional onshore OEL's were later granted to Esso, Safrap, and Great Basins. (See Table 2.1 for identification of these petroleum companies.) Furthermore, in 1960 Nigeria divided its offshore continental shelf into twelve blocks of about 1,000 square miles each. Ten of these blocks were taken up in 1961—four by Shell-BP, two by Gulf, two by Mobil, and two by Texas Overseas. The last-mentioned made Nigeria's initial offshore oil strike in 1963. Of the remaining two offshore blocks, Gulf took one in 1964 and Union obtained the other in 1967. Half of each of the original ten offshore concessions was relinquished in November 1968. The government accepted bids for new concessions in these areas during the first half of 1970.

Factors Influencing Oil Company Operations in Nigeria

A wide variety of economic and political factors influence oil company operations in Nigeria. One of the most important of these, the industry's financial arrangements with the Nigerian government, is sufficiently significant and complex to warrant very detailed discussion below and in Appendix B. Two physical advantages with economic importance are the location and the quality of Nigerian crude. Advantages of location include both international shipping charges and security of supply routes. As evidenced in Table 2.2, Western Europe (and especially the United Kingdom) is by far the most important market for Nigerian petroleum exports. With the closing of the Suez Canal and the constant threat of imminent upheaval in the Middle East, Nigerian crude is indeed favorably located, though not quite so auspiciously as that of its North African competitors.* In addition, Nigeria has relatively good access from a competitive point of view to its secondary markets—North America (United States and Canada) and South America (Argentina, Brazil, and Uruguay)—though less so than Venezuela does in North America.

* It is estimated that Nigeria's transport advantage over Persian Gulf crude destined for Western Europe amounted to about $0.27 per barrel before the closing of the Suez and currently runs about $0.47 per barrel.

TABLE 2.2. EXPORTS OF CRUDE PETROLEUM FROM NIGERIA, 1966

Country of destination	Volume (long tons)	Value (million £ Nigerian)
Western Europe:		
United Kingdom	7,373,251	36.3
Western Germany	2,678,557	12.9
France	2,221,384	10.7
Netherlands	556,721	2.7
Sweden	422,433	1.9
Denmark	403,851	1.8
Belgium	333,866	1.7
Norway	46,342	0.2
Spain	28,676	0.1
Subtotal	14,065,081	68.3
North America:		
Canada	1,303,076	5.8
United States	1,140,220	5.7
Curacao	328,438	1.7
West Indies	120,397	0.6
Puerto Rico	30,730	0.1
Subtotal	2,922,861	13.9
South America:		
Argentina	1,269,513	6.4
Brazil	371,857	1.7
Uruguay	200,783	0.9
Subtotal	1,842,153	9.0
Africa:		
Ivory Coast	30,085	0.1
Morocco	17,801	0.1
Subtotal	47,886	0.2
Australia	67,393	0.3

Unpublished data from the Nigerian Federal Office of Statistics. The production year 1966 is the last one unaffected by civil war interruptions.

Advantages of quality associated with Nigerian crude derive in large part from its relatively low sulfur content (0.3 per cent for API 34.6° crude oil).* This will become increasingly important as legislation concerning air pollution is passed in consuming coun-

* Nigerian crude is also attractive on the Western European market because of its relatively high fuel yield and excellent blending properties. Its relatively low kerosene yield makes it somewhat unattractive for the much smaller West African market. Average characteristics of Nigerian crude oil are as follows: specific gravity, 34.6° API; simple distillation ratios by percentage of weight, gasolines 27.1 per cent, middle distillates 41.1 per cent, residue 31.8 per cent; and residue characteristics, 0.30 per cent sulfur, pour point 37.

tries. There is technology available to remove sulfur from high sulfur crudes. But the world price of sulfur must rise even higher than recent levels to make the eventual joint production of petroleum products and sulfur more profitable than the production of petroleum products alone from a crude with low sulfur content. And even then the removal of sulfur from crude oil on a large scale would have a depressing influence on the world sulfur price, tending to shift the advantage back to the producers of low sulfur crude. Nigeria is not alone in possessing a "sweet" (low sulfur content) crude. Other sources include Libya, Algeria, and Indonesia. But with these exceptions all other major crude petroleum producers (except the United States and the USSR) have relatively "sour" (high sulfur content) crudes. Unless high sulfur prices induce a major cost-reducing technological change in sulfur extraction from crude oil, the "sweet" quality of Nigerian crude should remain an important advantage for the production of oil in Nigeria.

In addition to these physical influences on oil company operations in Nigeria, the industry has been directly affected by its relationships with government officials at several levels. Since Nigeria is the first tropical African country to develop its oil to any great extent, it is not at all surprising that expatriate officials of the oil companies, whose former experience in less developed areas was almost entirely confined to the Middle East, Venezuela, and Indonesia, would have a certain degree of difficulty in understanding some Nigerian ways of doing things. Conversely, Nigerian government officials, initially unaware of many of the complexities surrounding the international petroleum industry, have combined this lack of oil experience with a very justifiable feeling of nationalistic caution regarding foreign exploitation of a wasting natural asset. An intermixture of these two groups of decision-makers was bound to have uneven consequences. Gradual understanding of the intricacies involved in each other's decision-making process has thus been accompanied by a certain amount of misunderstanding and distrust of the other side's motives.

A few examples should suffice to establish this point.* Produc-

* The following incidents are based on accounts given directly to the author by officials of the various petroleum companies involved.

tion companies naturally must obtain the government's permission to construct oil pipelines. But one company once had to postpone production increases for over eighteen months while waiting for the government to approve its pipeline application leases. The difficulty apparently arose from a misunderstanding about which pipeline system design would minimize costs. Yet given the provisions of the profits tax legislation (explained below), both sides had an incentive to keep costs down to a minimum.

A second example involves the granting of expatriate quotas. The Nigerian government correctly desires to do all it can so that the oil industry's operations will train unskilled and semiskilled Nigerians and employ highly educated Nigerians. In order to do this, it is necessary to limit each oil company's use of foreign employees. But in the mid-1960's the quota policy was abused by both sides. Some of the companies applied for expatriate quotas far in excess of their actual needs in hopes of receiving at least their approximate requirements. This caused the government to become unusually restrictive in processing requests. Both sides had problems with the semantics of quota applications. The industry often did not make clear just what function an applicant was to undertake, while the government often did not appreciate the need for very highly specialized workers. At one point Nigerian immigration officials turned down an application for three "welders" because there were several unemployed Nigerian welders who they felt could handle the job. Furthermore, the applying company had specially trained three American welders, each with more than fifteen years of experience in specialized applications, for their highly skilled jobs in Nigeria.

A final example of misunderstanding between industry and government concerns the payment of import duties on certain types of equipment that the petroleum industry uses in its operations. The amount of taxes paid is not the major issue here, for the companies can use these import duty payments to offset petroleum profits tax liabilities (though the government would gain slightly by having its tax payments earlier). But assiduous or inexperienced customs officials have sometimes delayed clearance of imported equipment or spare parts at great cost to the companies involved.

In one instance a drilling company airfreighted a key part for one of its drilling rigs into Nigeria only to have it sit in customs for six weeks awaiting clearance. The rig made idle by this delay was renting for $10,000 a day. In all three of these instances, errors in judgment resulted ultimately in costs to both the government and the companies involved.

Financial Arrangements with the Nigerian Government

The financial arrangements under which the petroleum companies operating in Nigeria agree to make payments to the Nigerian government are contained partly in various Nigerian laws and decrees and partly in the covenants that each company has with the government.* Though these secret covenants reportedly differ somewhat in detail, the major financial arrangements apply uniformly to all companies. It is thus meaningful to speak of tax terms for the industry as a whole. The methodology that one can employ to calculate the annual payments that each company makes to government is technically complicated and analytically of limited interest. Much of this discussion is therefore relegated to Appendix B. Discussion here incorporates only those details that are directly pertinent to an understanding of the economic and political results of the petroleum industry's financial arrangements in Nigeria.

The first of two important and interrelated results derives from the economic and psychological effects on decisions by each company's head office with respect to production levels. The cost-price structure, in Nigeria as elsewhere, causes tax considerations to dominate cost factors in the international petroleum industry's production operations.† Naturally any change in tax terms will change the calculation of tax-paid cost per barrel for each produc-

---

* See Appendix B for a more detailed discussion of the technical aspects of this topic. Except in instances noted, the sources of the information set forth in the remainder of this chapter are the various pieces of Nigerian legislation listed at the beginning of Appendix B.

† Indications are that the costs of producing petroleum in Nigeria are steadily decreasing as a result of increased familiarity with Nigerian conditions and of technological innovations. Between 1961 and 1965, the average time required by Shell-BP to drill 10,000 feet was reduced from sixteen days to nine days. A major technological innovation, developed by Shell-BP, injected a plastic sub-

tion company. Moreover, the manner in which the change is negotiated may have direct relevance for company expectations of possible future changes. Second, in addition to this indirect effect on production levels, the financial arrangements clearly have a direct influence on petroleum industry payments to the Nigerian government. This is the vehicle through which the industry can have its greatest impact on the Nigerian economy.

Original financial arrangements. Under the Petroleum Profits Tax Ordinance of 1959, the Nigerian government instituted the standard fifty/fifty arrangement whereby the government and the petroleum company in question shared that company's profits equally.* More specifically, under these arrangements the producing company applied its realized price to the volume of crude oil produced and exported to calculate gross proceeds. This valuation of output served as the basis for figuring royalties and taxes on profits. Rentals and other minor taxes were calculated separately. Capital allowances (i.e. depreciation) were generated according to legislation providing for accelerated write-offs.

The calculation of taxes on profits then proceeded as follows. (See Appendix B for an algebraic explanation of the petroleum industry's financial arrangements with the Nigerian government.) All expenses (operating costs and intangible appraisal or development drilling costs) were subtracted from gross proceeds. A stipulation was made that at least 15 per cent of this difference had to be paid as taxes on profits no matter what amount of capital allowances was available. The allowable amount of capital allowances was then subtracted to arrive at the chargeable profit. Of this figure half was paid to the government. But included in this total payment were the royalties, rentals, and other minor taxes paid, for which no service was performed; all of these were thus used as offsets against profits tax. In no instance would the company's total

---

stance into oil well formations to prevent sand influx. An important negative cost factor is that Nigerian oil wells and fields are small and scattered, especially relative to the Middle East standards. See *The Shell-BP Story*, pp. 13–16.

* The term profits is employed throughout this discussion to accord with the legal terminology. From an economist's point of view, what the company and government share equally is mainly economic rent, not profits.

tax obligation to the government exceed 50 per cent of the chargeable profit.

Alterations in the financial arrangements. In late 1966 and early 1967 the Nigerian government instituted two important changes in its financial arrangements with the petroleum industry. These changes maintained the earlier framework—e.g., retained the fifty-fifty profit-sharing arrangement—but altered some of the provisions within it. The federal military government's Decree Number 65 of 1966 modified the capital allowance legislation affecting the oil industry among others (see Appendix B, pp. 180–81). Prior to this decree, there had been separate legislation pertaining to petroleum companies' depreciation allowances. It is widely suspected within the oil industry that the government mistakenly forgot to exclude the petroleum companies in the new legislation and then refused to correct this oversight. In any event, the rate at which the companies are allowed to depreciate their capitalized investment was cut nearly in half. Slower generation and use of capital allowances means larger chargeable profits, and thus the stream of profits taxes over time is shifted forward to Nigeria's advantage.

The second and much more important alteration was contained in Decree Number 1 of 1967. In this decree the federal military government imposed OPEC (Organization of Petroleum Exporting Countries) terms on the companies operating in Nigeria.* Before describing these terms, it might be instructive to review the evolution of this change. In each of the company covenants with the Nigerian government there reportedly exists a provision usually termed the most-favored-nation clause. In this provision the companies operating in Nigeria promise that they will nego-

* There is one important difference between OPEC terms and the new Nigerian arrangements. Companies operating in OPEC countries have agreed to phase out the 6½ per cent "percentage allowance" discount off posted price by 1972 and the "gravity allowance" discount by 1975. The Nigerian government has not yet negotiated these concessions. In spite of this difference, the new arrangements in Nigeria are referred to as "OPEC terms" for convenience in the ensuing discussion.

tiate with and, by implication, give terms to the Nigerian government equal to the most favorable terms accorded to any other government on the continent of Africa or in the Middle East. In late 1965, the Libyan government exacted further tax concessions from the petroleum companies operating in Libya. Once the companies that are active in both Libya and Nigeria acceded to the Libyan demands, it was only a question of time when the Nigerian government would decide to ask for equal terms. It should be pointed out that Nigeria has not been, and still is not, a member of OPEC. But under the terms of the company covenants it is not necessary for Nigeria to join OPEC in order to receive treatment equal to that accorded to Libya or to any other African or Middle Eastern country.

Basically, the change to OPEC terms involves two things. First, the companies must agree to set a posted price. What this means is that the companies agree on a procedure to establish a fair price for exported oil on which petroleum profits taxes as well as petroleum royalties will be calculated. De facto, however, once one company sets a posted price, each of the others eventually must set the same price, because there reportedly exists a most-favored-company clause in each of the company covenants with the Nigerian government stating that no one company will receive better treatment than any of the others. The second new provision included in OPEC terms is the "expensing" of royalties, i.e. the treatment of royalties as expenses rather than tax offsets. Under the previous arrangements, petroleum companies could use the royalties paid to government as a 100 per cent offset against petroleum profits tax liabilities. Now, however, royalties no longer are tax offsets but are treated as expenses. The net impact of this change is that Nigeria will receive additional profits taxes amounting to 50 per cent of the value of royalties.

The manner in which OPEC terms were introduced into Nigeria is a prime demonstration of the fact that the companies operating in Nigeria are not at all the solid monolith that some casual observers claim them to be. Indeed, in Nigeria, as elsewhere, there is con-

siderable rivalry among oil-producing companies. Shell-BP, the company with by far the longest history, the largest operation, and the brightest future in Nigeria, understandably takes the lead among the exploring and producing companies. But several of the other companies, especially certain American-owned ones, have not always been pleased with the ways in which Shell-BP has chosen to exert its dominance. These dissatisfactions surfaced explosively over the issue of OPEC terms.

According to the permanent secretary of the Federal Nigerian Ministry of Finance, Shell-BP supplied the initiative for the Nigerian emulation of the Libyan changes by informing the Ministry of its desire to afford the better terms to Nigeria.[8] A committee established in May 1966 by the first military government considered Shell-BP's proposal, but action was postponed to the end of that year because of the change of government in July. One part of the proposal involved making the new terms retroactive to the beginning of the year in which the change took place. In its haste to seal the agreement before the end of 1966, the Nigerian government only had time to negotiate with the three companies that were at that time in production—Shell-BP, Gulf, and Safrap. Gulf initially declined to agree, while Safrap asked for additional time to study the matter. Both of these companies and nearly all of the other companies were reportedly offended at the manner in which the government and Shell-BP handled this matter.

Of course, much more than proprieties was at stake. The greatly increased effective tax rates affected the financial position of all of the follower companies. The timing of the change was crucial. Though all of the companies realized that they would eventually have to switch to OPEC terms in Nigeria, several hoped that this change would be delayed until they had achieved stronger financial positions. As a result, an extensive period of ill feeling ensued among certain of the managers of the Nigerian companies, perhaps reflected most directly by the cessation for over a year of the monthly luncheon meeting of the oil producers' section of the Lagos Chamber of Commerce. (By early 1969, relations among companies had considerably improved.)

After a long delay, in November 1967 a posted price of $2.17 per barrel for API 34° crude was announced by Shell-BP.* In light of the most-favored-company clauses, it was apparent that all of the other companies sooner or later would have to fall into line. At present the other companies have either agreed to this posted price or are still in the process of negotiating with the government. The Shell-BP posting is effectively a price of $1.948 per barrel; from the initial level of $2.17 one must deduct $0.06 per barrel for harbor dues and then take the percentage allowance ($0.137), gravity differential ($0.02), and marketing allowance ($0.005) discounts from the residual figure of $2.11 per barrel. The OPEC countries have recently renegotiated the question of discounts. The Nigerian government may decide to follow suit in the not-too-distant future. Until this occurs, however, an effective posted price of about $1.95 per barrel for 34° Nigerian crude can safely be assumed to apply uniformly.

When a company calculates its payments to government on the basis of a realized price, it has the prerogative of stating what realization price it has obtained. In the recent past Shell-BP has used a price for tax purposes of about $1.95 per barrel. In effect, then, Shell-BP posted its previous realized price. This action forces other companies that were using, or were planning to use, a lower realized price to pay higher taxes. The belief that the price posted by Shell-BP is too high has been the cause of considerable acrimony in the industry.

In addition to these two very important changes in financial arrangements, the Nigerian government instituted a third set of potential alterations with the Companies' Decree, issued in October 1968. In essence this decree forced all companies operating in Nigeria but incorporated abroad to become Nigerian corporations. As with the new legislation on capital allowances, petroleum companies were treated no differently from other private foreign con-

---

* *Platt's Oilgram News Service*, November 7, 1967. Technically, the parent companies, Shell and BP, actually posted the price of $2.17 for crude in the API range 34–34.9° (less harbor dues). There is a $0.02 variation for each whole degree of gravity; for example, the posted price for 27.0–27.9° API crude is $2.03.

cerns, in spite of the very special nature of their operation. No immediate economic advantages associated with petroleum accrue to Nigeria from the enactment of the Companies' Decree. And although it is too early to be conclusive, the long-run effects of this decree on the oil industry will probably be negligible. Naturally only those companies not already incorporated in Nigeria were affected. But each affected company spent much time in legal complexities to assure that Nigerian incorporation would not impair existing tax advantages, especially those concerning the carrying forward of tax credits in Nigeria, the retention of United States advantages, and the avoidance of a transfer tax imposed by the U.S. Internal Revenue Service (the latter two would affect American companies only). On the other hand, short-run costs associated with this decree are not insignificant. The decree caused one company to postpone going into production for several months until all legal details were settled. And the waste of perhaps four to six months of executives' time has been accompanied by a general worsening of company-government relations.

Comparison of original and current financial arrangements. A simplified example follows, showing the possible impact of OPEC terms, the most significant change in financial arrangements. Assume under the prior terms that a company exports one barrel of oil with a realized price of $1.80. Assume further that all costs of production amount to $0.50 and that the company has capital allowances of $0.20. The total chargeable profit would then be $1.10, i.e. $1.80 less $0.50, less $0.20. Under the old arrangements, 50 per cent of this amount went to the Nigerian government and 50 per cent was retained as profits by the company. The government's $0.55 comprised the total tax obligation, since royalties, rentals, and other tax offsets were all used to decrease the tax liability on the company's petroleum profits. Assume that the royalty that the company had to pay on its one barrel of oil amounted to 10 per cent of the realized price, or $0.18, and further, that rentals and other tax offsets amounted to $0.08. Under the old legislation this would mean that the total payments to government, which must total $0.55, would then be made up of $0.18 royalty, $0.08 rentals and other tax offsets, and $0.29 profits tax.

The introduction of OPEC terms changes the situation considerably. First, the company must set its posted price. Assume that this is $1.95 for the one barrel of oil that the company exports. The costs and capital allowances remain unchanged at $0.50 and $0.20, respectively. Now, however, the royalty, which again is assumed to be 10 per cent of gross proceeds, must be calculated on the basis of the posted price; the royalty thus amounts to 10 per cent of $1.95, or $0.20. Furthermore, this royalty cannot be used as a tax offset but must be charged to expenses. The $0.20 royalty is therefore subtracted from gross proceeds along with the $0.50 costs and the $0.20 capital allowance, leaving a chargeable profit of $1.05. And, as before, 50 per cent of this amount goes to the Nigerian government and 50 per cent is retained as profits by the oil company. The company still is able to offset its $0.08 worth of rentals and other tax offsets against its tax liability on petroleum profits of $0.53. But now the total payments to government equal $0.73, made up of $0.45 profits tax, $0.08 rentals and tax offsets, and $0.20 royalties.

Finally, it is instructive to compare the total payments to government with the company's retained profit in each instance. In the first example, assuming prior financial arrangements, the company retains $0.55 and the government receives $0.55. In the second example the company still receives a realization of $1.80 for its barrel of oil exported. From this $1.80 actually received it has costs of $0.50 and capital allowances of $0.20, leaving $1.10. And of this $1.10 a total of $0.73 is paid to government. Profits now are only $0.37. The effective split is no longer 50–50 as under the old legislation, but 66–34 in favor of the Nigerian government. Naturally, the split is less favorable for the government if a higher realization price is attained. In fact, this entire argument hinges on the level of the price actually realized by the company when it internally transfers crude produced in Nigeria. In Chapter 10 an analysis of the effect of the changed terms on projected government revenues is presented, employing a constellation of different realization prices. The lessons contained in the hypothetical discussion of this section are pertinent to an understanding of this key policy issue.

Determination of Nigerian Petroleum Production

Location and quality factors together with local administrative influences are important in the determination of levels of Nigerian production. But the various head offices determine annual levels of production for their Nigerian production subsidiaries primarily on the basis of tax-paid cost per barrel of crude, adjusted for transport differentials and crude quality differences, with large and undefinable weights given as well to security and diversity of supply, political stability, considerations in other producing countries, and a host of other extra-economic factors. A production company normally arrives at its tax-paid cost per barrel of crude for each prospective investment project by assuming a time frame that ends with the expiration of its concession license and then comparing a discounted stream of projected cash outlays with a discounted stream of expected incomes.*

The head office of the entire integrated operation periodically reassesses its global marketing position and then reallocates production to its various producing areas by choosing the most promising investment alternatives. The subsidiary production company, in Nigeria as elsewhere in the petroleum-producing world, generally finds itself in the position of trying to convince its head office to accept increasingly higher minimum levels of production in its desire to produce as much as possible of the oil that it has worked so hard to find. But head offices plan globally in their efforts to balance supply and demand and often must restrain production subsidiaries from expanding as rapidly as they might like. The upshot of this, especially for a country like Nigeria in which production is just beginning to accelerate, is that the production companies are unable to predict with any great assurance what levels of output they might be able to attain. And this makes accurate medium-term planning as hazardous for the Nigerian government as for the production subsidiaries themselves.

* Nigerian Oil Mining Leases run for 30 years onshore and 40 years offshore. But because pipeline licenses run only for 20 years, some oil company officials use this shorter time span in calculating tax-paid cost per barrel. Note that for new concessions the period for OML's will be 20 years according to the provisions of the Petroleum Decree 1969.

CHAPTER THREE

Recent Trends in Nigerian Economic Development

This chapter is a brief summary of what others have said concerning growth in pre-civil war Nigeria.[1] This account will set the stage for later analysis of interactions between the petroleum sector and other sectors in the Nigerian economy. The choice of topics is dictated by their importance in recent Nigerian development and their relevance for a comparative understanding of petroleum's recent and likely future role in this process.

Recent Growth Trends

Nigeria initially experienced rapid growth of exports in the first three decades of this century, following the imposition of British colonial rule. Between 1900 and 1929 export value grew sevenfold while export volume increased five times; these increases translate into compounded annual growth rates of 7 per cent and 5½ per cent, respectively.[2] This expansion was based on palm produce, rubber, groundnuts (peanuts), cocoa, and cotton—the same commodities that have been and still remain Nigeria's major agricultural exports. Then, as now, these crops were produced by farmers with small landholdings employing a traditional technology.*
After the onset of the Great Depression in 1929, the Nigerian economy entered a long period of stagnation that ended only after World War II. Evidence of this stagnation is shown in the fact that between 1929 and 1945 export volume grew by only 20 per

---

* The term traditional implies export production on a small scale with a simple technology and some orientation toward subsistence production. See Gerald K. Helleiner, *Peasant Agriculture, Government, and Economic Growth in Nigeria* (Homewood, Ill.: Richard D. Irwin, 1966), p. xii.

TABLE 3.1. SECTORAL CONTRIBUTION TO GDP (AT FACTOR COST),
SELECTED YEARS 1950–68
(*Per cent*)

Item	1950	1954	1958	1962	1963	1964	1965[a]	1966[a]	1967[a]	1968[a]
Agriculture, livestock	64	62	66	59	57	52	51	50	49	50
Fishing, forestry	3	4	3	4	3	4	4	4	4	5
Petroleum	...[b]	...[b]	...[b]	1	1	2	3	4	3	1
Manufacturing, public utilities	1	1	3	4	5	5	42	42	43	44
Other	33	34	28	33	35	37				
Total[c]	101	101	100	101	101	100	100	100	99	100

Based on Appendix A, Table A.1.
[a] Unofficial estimates.
[b] Negligible.
[c] Differences from 100 are due to rounding.

cent (over the entire period), or at approximately the same rate as population and domestic output might have been growing.[3]

After 1945, aggregate indicators of Nigerian economic progress began a steady upward trend. On the basis of official figures, Gross Domestic Product (GDP) at factor cost in constant 1962 prices grew at 4.1 per cent between 1950 and 1957 and 5.7 per cent between 1958 and 1966.[4] Appendix Table A.1 contains a summary of the most import economic indicators for all of the years for which estimates exist.[5] Some of the information contained in this detailed table is summarized in Table 3.1, where the percentage sectoral contributions to GDP for selected years are displayed. Agricultural value added was dominant throughout the postwar period, but fell from 64 per cent in 1950 to 52 per cent in 1964 (and to an estimated 50 per cent in 1968). In recent years, the ratio of value added from agricultural exports to total agricultural value added has been about .15. And it has been estimated that roughly 25 per cent of total agricultural output has been traded internally.[6] These estimates indicate that subsistence agricultural production has lately been responsible for perhaps 60 per cent of total agricultural value added and thus about 30 per cent of GDP.*

* To the extent that subsistence production is undervalued in GDP accounting, total GDP is understated whereas growth of GDP is overstated.

No one disputes the fact that the Nigerian economy has been growing adequately during the past two decades. Recent studies, however, evidence some disagreement about which elements in the economy might best be credited as prime movers in this process. For example, both Helleiner and Lewis give preeminence to the role of traditional agricultural exports.[7] Helleiner, however, stresses the positive part played by the Nigerian government, whereas Lewis minimizes this and gives second mention to manufacturing. These three influences are now summarized.

Agricultural Exports

Table 3.2 shows that Nigeria has a reasonably diversified mix of agricultural exports that tends to moderate adverse effects of wide commodity price fluctuations.[8] Major agricultural exports have grown in value at an annual rate of 4.6 per cent between 1950 and 1966 and 3.8 per cent between 1958 and 1966.*

The factors underlying this substantial growth of agricultural exports are complex. Lewis documents the major incentives as favorable prices (see below for qualifications), the growth of transport facilities, and technological changes.[9] But Helleiner sees the farmers' production responses as due primarily to extensive use of previously unused factors, including both land, of which Nigeria has had and still has an abundant supply in most areas, and labor, from a growing population and decreased leisure, though productivity has increased with some crops.[10] Technological improvements have increased productivity in cocoa and, to a lesser degree, cotton and palm products, and have been most lacking in rubber and, until more recently, groundnuts.[11]

Manufacturing

Nigerian industry has recently (1950–63) grown very rapidly—15 per cent per annum considering only large-scale manufacturing (ten or more employees) and 10 per cent per annum including both large- and small-scale—but from a very small base.[12] By 1963

---
* These rates may be somewhat understated by the choice of 1966 as a terminal year owing to the poor cocoa performance in that year. Consideration of 1967 and 1968 is omitted because of unusual adverse effects caused by the civil war.

TABLE 3.2. COMPOSITION OF DOMESTIC EXPORTS, SELECTED YEARS 1900–1968

Commodity	1900	1929	1946	1950	1954	1958	1962	1963	1964	1965	1966	1967	1968
					PER CENT OF VALUE								
Cocoa	—	13	16	22	27	20	20	18	19	16	10	23	25
Groundnuts (nuts, cake, and oil)	—	14	24	17	23	23	24	25	22	20	20	21	25
Palm products[a]	82	46	26	33	16	25	16	16	15	15	14	6	7
Cotton (raw and seed)	—	3	2	3	5	6	4	6	4	2	3	4	3
Rubber	10	1	6	3	2	6	7	6	6	4	4	3	3
Subtotal	92	77	74	78	73	80	71	71	66	57	51	57	63
Petroleum[b]	—	—	—	—	—	1	10	11	15	26	33	30	18
Other exports	8	23	26	22	27	19	19	18	19	17	16	13	19
Total domestic exports	100	100	100	100	100	100	100	100	100	100	100	100	100
					VALUE OF EXPORTS (million £ Nigerian)								
Domestic exports	1.9	17.6	23.7	88.5	146.2	132.8	164.0	184.9	210.2	263.3	278.7	238.1	206.5
Reexports	—	0.2	0.9	1.7	3.3	2.8	4.5	4.8	4.2	5.0	5.4	3.7	4.6
Total exports	1.9	17.8	24.6	90.2	149.5	135.6	168.5	189.7	214.4	268.3	284.1	241.8	211.1

Data from Helleiner, *Peasant Agriculture*, pp. 492–93, 530–31; Nigeria, Federal Office of Statistics, *Review of External Trade, Nigeria 1968* (April 1969), pp. 8 and 11; and Nigeria [15], p. 16. (Frequently cited government publications are referred to by bracketed numbers. Full citations are listed in the Selected Bibliography, pp. 219–24, under Nigeria.)

[a] Palm oil, palm kernels, palm kernel oil, and palm kernel cake.
[b] The comparison of gross petroleum earnings with non-oil export earnings overstates the actual contribution of petroleum to export proceeds. See pp. 74–77.

the manufacturing sector still accounted for less than 8 per cent of GDP.* Industrial activity has been heavily concentrated in import-substituting ventures and in the processing of agricultural commodities for export.[13] As Kilby notes: "Industrial production in Nigeria ... exhibits wide diversity in terms of the degree of specialization and division of labour, technology, factor proportions, the quality of raw material input and product finish, the character of markets being served, and entrepreneurial organization."[14] Lewis estimates that Nigerian national income has benefited from about 30 per cent of value added in large-scale industry, a low percentage but one that is not at all unusual in developing countries.[15] Small-scale industry has received relatively little attention from empirical analysts.[16] Several observers have noted the potential advantages in the encouragement of small industry, notably entrepreneurial development and production of simple agricultural implements.[17] Manufacturing will play an increasing role in Nigerian development, though it will be some time before its impact approaches that of either agricultural or petroleum exports.

Government

The role of government in economic development consists basically of policy-making and allocation of public funds.[18] Government economic policy in Nigeria prior to the civil war was oriented toward a free market, and incentives were established to encourage private investment. In public investment the Nigerian federal and regional governments concentrated their efforts in four areas. First, they invested quite heavily in infrastructure, especially transportation and communication facilities, mostly with good economic results.[19] Second, they undertook four types of agricultural development programs—plantations, tree subsidies, research and extension, and land settlements—with mixed economic success.[20]

---

* W. Arthur Lewis, *Reflections on Nigeria's Economic Growth* (Paris: Development Centre of the Organisation for Economic Co-operation and Development, 1967), p. 23. Estimates of the manufacturing sector's share of GDP generally run much lower than this; see, for example, the official estimate contained in Table 3.1 that shows only about 5 per cent for manufacturing and public utilities. The higher figure reflects reasonable adjustments by Lewis.

Third, they spent large amounts, relative to total budget, on education that had likely long-term benefits, with both positive and adverse short-range effects.[21] And fourth, they participated directly in industrial ventures, with very uneven economic outcomes.[22]

There is some debate over the importance of the Nigerian federal and regional governments in stimulating economic growth. Helleiner, for example, feels that "the government has constituted the most important source of initiative and the most dynamic element in recent Nigerian economic experience."[23] But Lewis strongly disagrees, stating that economic development "resulted more from the adaptability of the farmers, the energy of the business community, and the resources of the country, than from anything that the governments were doing."[24] In any event, government policy will surely become increasingly important in the future development of Nigeria.

Special Problem Areas

Recently there has been growing concern over the possibility that the Nigerian governments and marketing boards might be squeezing the agricultural export sector too hard.[25] Tables 3.3 and 3.4 together produce some convincing evidence in support of this argument. Table 3.3 shows that between 1957 and 1963 F.O.B. export prices for most of Nigeria's major agricultural exports rose, peaked, and then returned to levels slightly below those of 1957. Concurrently, producer prices of all marketing board commodities (save groundnuts) followed a steady downward trend, so that at the end of the period Nigerian producers were squeezed relatively more than at the beginning.

Table 3.4 indicates that in this same period the ratio of producer price to food price moved so as to establish an incentive to switch from some export commodities into domestic food crops. This incentive has undoubtedly been responsible for the decline in exports of palm products in the 1960's. With producer prices between 50 and 70 per cent of F.O.B. export prices, there is clearly sufficient scope for substantial decreases in export duties, sales taxes, and/or marketing boards' profits.[26] Relief for the small farm exporters

TABLE 3.3. F.O.B. EXPORT PRICES AND PRODUCER PRICES FOR MAJOR NIGERIAN AGRICULTURAL EXPORTS, 1957–63

Commodity	1957 (£ Nigerian per long ton)	Indexes 1957 = 100					
		1957	1958	1959	1960	1961	1962	1963

F.O.B. EXPORT PRICES

Commodity	1957	1957	1958	1959	1960	1961	1962	1963
Cocoa	193	100	159	138	124	95	89	98
Palm Oil	83	100	89	90	92	96	91	91
Palm kernels	44	100	105	137	141	109	104	118
Seed cotton	89	100	92	80	88	92	100	96
Rubber	176	100	105	124	141	113	108	106
Groundnuts	67	100	79	83	103	98	91	89

PRODUCER PRICES

Commodity	1957	1957	1958	1959	1960	1961	1962	1963
Cocoa	146	100	100	107	106	66	68	72
Palm oil	39	100	110	110	100	92	77	77
Palm kernels	30	100	100	100	97	94	80	80
Seed cotton	56	100	100	102	102	89	80	80
Rubber[a]	176	100	105	124	141	113	108	106
Groundnuts	32	100	89	137	140	130	121	121

Data from Nigeria [10], pp. 92–94; and Lewis, p. 21.
[a] F.O.B. export price; there is no marketing board or producer price for rubber.

TABLE 3.4. RATIOS OF PRODUCER PRICE TO FOOD PRICE
(1957 = 100)

Commodity	1957	1958	1959	1960	1961	1962	1963
Cocoa	100	122	120	110	57	50	70
Palm oil	100	121	117	101	106	82	72
Palm kernels	100	110	106	98	107	85	74
Seed cotton	100	93	100	127	106	74	85
Rubber[a]	100	128	139	146	98	79	103
Groundnuts	100	83	134	175	154	112	128

Data from Lewis, p. 21. Producer prices are those declared for the most common grade by the marketing board of the region of principal production. The food prices are weighted averages of retail prices of Western Region roots (for cocoa and rubber), Eastern Region roots (for palm oil and palm kernels), and Northern Region cereals (for cotton and groundnuts).
[a] Ratio of F.O.B. export price to food price; there is no marketing board or producer price for rubber.

could take the form of increased direct incomes (via lower taxes) or subsidized agricultural inputs.[27] Policies that provide the proper incentives could greatly enhance the prospects for future Nigerian agricultural development.

The combination of the decrease in prices paid to farmers and the rapid rise in urban wages has been partly responsible for a

second problem—the large and increasing amount of urban unemployment.[28] This problem has been exacerbated by the migration into the cities of semi-educated young people who disdain rural employment.[29] This is, of course, partly the result of the heavy emphasis that southern Nigerian regional governments have placed on primary education at the expense of secondary education.[30] Currently the unemployment problem is alleviated by military employment. But, as Lewis argues, the long-run solution to urban unemployment will most easily be found in expansion of the services sector. This expansion must be based on accelerated growth of Nigeria's prime movers—petroleum exports, agricultural exports, and manufacturing.[31]

In summary, Nigeria faces many of the general unresolved questions of development that confront any low income economy. These include allocational questions concerning the long and the short run, direct versus indirect assistance to productive sectors, domestic or foreign ownership, the need for physical plant as well as health and educational facilities, and the proper degree of government intervention in industrialization. Most important, answers to these questions must be found within a framework of potential political instability and regional conflicts.

CHAPTER FOUR

Foreign Investment in a Less Developed Economy

In the framework designed for this study, it is assumed that there is an opportunity for direct investment to be financed by private foreign resources and that this investment does not require restrictions on imports.* Primary emphasis is given to the effects of the foreign investment on the less developed economy. Such questions as the rate of return on investment are ignored. It is also assumed that sufficient local or external demand exists to attract the private foreign investment. The framework is thus constructed to place major concern on influences of supply rather than demand.[1]

The Direct Contribution

The first problem encountered in attempting to evaluate the effect of any economic activity is the determination of a suitable measure of economic welfare. For this purpose most economists tend to use consumption or growth of consumption, sometimes approximated by national income or its growth, and usually considered per capita. In analyses of less developed economies, especially those with very large and rapidly growing populations, econ-

---

* Direct private foreign investment is defined as including foreign-financed projects in less developed economies in which the foreign investor retains control of management decisions, normally by having at least 51 per cent equity. All other private foreign capital flowing to less developed countries can be termed portfolio investment. The framework employed in this study pertains to direct flows, though in large part it could apply as well to portfolio investment. To apply this framework to import-substituting industries financed by private foreign investment, it would be necessary to consider explicitly several issues, such as level of protection, that can conveniently be disregarded in dealing with export industries.

omists are increasingly using employment of local labor or the growth of this indicator as an additional criterion of welfare. Other very important criteria are the distribution of income and several extra-economic influences, including political and sociopsychological effects. In this study primary emphasis is given to the direct and indirect contributions to national income associated with private foreign investment, though more than passing attention is paid to political and other influences.

Private foreign investment will generally provide annual streams of direct contributions as well as various indirect contributions to national income in a less developed country. The direct contribution involves local payments to factors of production made by the newly established foreign-owned industry. This direct contribution of private foreign investment must be measured net of the social opportunity costs of domestic resources employed. Domestic inputs complementary to the foreign investment do not have a zero opportunity cost unless they are unemployable elsewhere. In order to find the net direct contribution of private foreign investment, for each local factor of production receiving a local payment one must calculate the difference between its total payment and the total payment that it could otherwise obtain in alternative employment in the local economy, i.e. its social opportunity cost, and then find the sum of these differences.* The net gain in any year can usefully be related to that year's total receipts of the industry created by the foreign investment. This ratio can be termed the annual net gain coefficient. If values of net gain from different

---

* The net gain concept is summarized in the equation

$$G = \sum_{i=1}^{m} (P_i - C_i) ,$$

where $G \equiv$ net gain from direct private foreign investment,
 $P_i \equiv$ total payment by the new industry to local factor i,
 $C_i \equiv$ social opportunity cost of local factor i,

and local factors, 1, ..., m are all employed by the new activity financed by the foreign investment. This formulation abstracts from the effects that might be associated with a change in the price of the output produced by the new industry. In addition, general price inflation is assumed not to exist or to be continually corrected for, i.e., the variables in the equation are measured in constant prices.

years are to be compared, it would, of course, be necessary to employ a discount factor, in this instance most appropriately a social discount factor.

Some of the information required to estimate the net gain from foreign investment and hence the net gain coefficient is difficult to obtain. The annual totals of receipts associated with the investment are known, of course, as are the amounts paid locally to factors of production. The elusive part of the procedure lies in the approximation of each factor's social opportunity cost. In a formal sense, one might handle this problem by constructing a mathematical programming framework and solving for the binding factor constraint(s) (see pp. 42–45). Then in theory all nonbinding factors would have a zero opportunity cost for the economy. But this approach is limited by restrictive underlying assumptions, especially the postulate of linearity and the assumption that the short-run supply of binding factors is completely inelastic. In the real world, opportunity costs are better estimated with a heavy input of intuition.

It seems clear that the opportunity cost of all imported goods and services will be at least equal to their entire value, if foreign exchange is scarce. Local materials, however, will have a lower opportunity cost to the extent that idle factors are used in their production. The opportunity cost of rents paid to the government for use of domestic natural resources will depend on the given set of circumstances. If the local government or domestic investors could immediately exploit the resources, then clearly a social opportunity cost is involved. If not, this cost is more difficult to approximate, since the future discounted flow of rental payments that will show up as gain derived from the foreign investment must be compared with the future discounted flow of returns associated with later governmental exploitation. In addition, any returns to the private foreign investment that are retained for use in the domestic economy rather than repatriated will generally have a zero opportunity cost to the economy. It is most convenient to consider retained earnings as new private foreign investment and hence to evaluate direct and indirect gains separately.

But how one should treat the opportunity costs of salaries and wages is a subject of great dispute. Salaries paid to skilled local employees as well as the local currency component of expatriate salaries might have a social opportunity cost virtually equal to the entire amount of the payment. This would be so to the extent that these employees could receive an equivalent marginal value product in alternative employment in the domestic economy. On the other hand, wages and salaries paid to semiskilled and especially unskilled labor almost surely do not have a full 100 per cent opportunity cost, for some of these laborers would be underemployed or unemployed in the absence of the foreign investment. It is difficult to estimate the true opportunity cost of unskilled labor. As a first approximation for most low income economies, it is probably greater than zero but closer to zero than to the going wage for unskilled labor.

Attraction and Creation of Factors of Production

Direct private foreign investment might result in two kinds of indirect contributions to growth of national income in a low income economy. One of these involves the attraction or creation of factors of production. Initially the industry established with the foreign investment attracts foreign capital, entrepreneurship, labor, and management. The creation of factors implies that in the process of its operation the new industry might contribute certain types of factors to the domestic economy, particularly foreign exchange, investment resources (especially if large payments to government occur), and the training of skilled labor and management. In addition, factors might be created through the discovery of new resources or the opening of frontiers.

Either foreign exchange, investment resources, or the country's absorptive capacity as shown by available skilled labor might at any given point in time constrain further increases in output. The foreign capital investment results in a flow of foreign exchange into the domestic economy. If the foreign investment occurs in an export industry or in an import substitution industry that creates a net saving of foreign exchange, then the industry will be a pro-

vider of this potentially important factor to other domestic sectors. In addition, the foreign capital inflow automatically translates into available investment resources for the domestic economy that presumably are used mainly to establish the new industry. If in addition to these investment resources there is private or public domestic saving generated by the operation of the new industry, then the foreign capital flow results in a net contribution of investment resources to the domestic economy. Finally, skilled laborers (including entrepreneurs and managers) might be trained in the new industry and then choose to move to other domestic industries. To the extent this occurs, the new industry supplies skilled labor to the local economy in amounts greater than its own requirements.* Since the new industry might be a net user rather than supplier of one or more of these factors and since some factors might be supplied temporarily and then withdrawn, all factor flows must be estimated on a net basis. These considerations also argue for use of a time horizon that coincides with the life of the foreign investment where possible.

The principal issue in measuring the importance of these factor contributions is the difficult task of identifying the limiting factor constraints and their accompanying shadow prices, i.e. the additional output obtainable from an extra unit of the binding factor. Low income economies are in general prevented from attaining faster growth of output because of the rigidity of supply. To the extent that only very limited factor substitution can occur in the short run, it is useful to search for the factor or factors of production whose scarcity limits growth of output during any given period of time. Once the binding factor is identified, one can measure the importance of additional contributions of this factor (supplied by the new industry or by any other source) by calculating associated increases in output, thereby estimating the factor's shadow price. For example, if foreign exchange is binding, one extra unit of this factor might allow the economy to achieve, say, five extra units of gross output. Clearly, if resources cannot be sub-

* The extent of training effects will depend partly on the skill- and industry-specificity of the jobs created by the new industry.

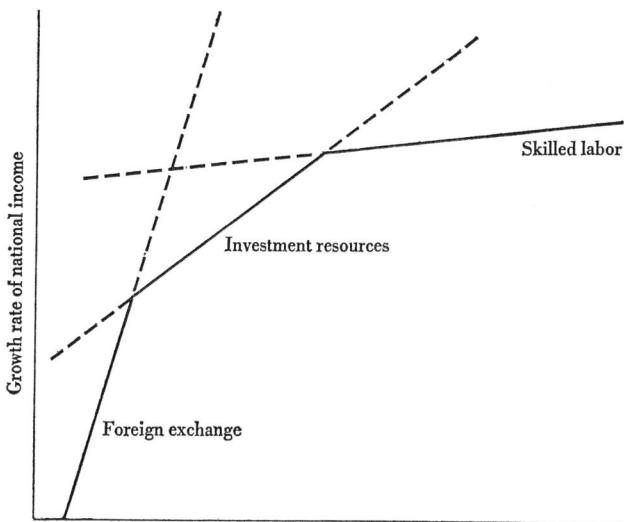

Demonstration of possible factor constraints. Solid line equals maximum growth rate of national income for any given value of ratio of foreign exchange to national income. Source: McKinnon, pp. 388–409; adopted by Roemer, p. 83.

stituted in the short run, additions of nonbinding factors will have no positive effect on output, although this could eventually cause shifts in the combination of inputs used.*

It is reasonable to suppose that growth of output in low income economies is constrained by the limited availability of foreign exchange, investment resources, or skilled labor. Other factors can be considered in excess supply (e.g. unskilled labor) or part of one of the three mentioned inputs (e.g. land in investment resources, entrepreneurial capacity in skilled labor). The figure shows these

---

* In theory it is possible that this particular type of model might be inappropriate. This would be so if an underdeveloped economy has relatively easy substitution among its sectors and each sector has more or less fixed proportions. If this were indeed the case, then a neoclassical aggregate production function would be a more reasonable approximation. This is an interesting theoretical possibility but one that is unlikely to occur in practice. For a related discussion see Michael Roemer, "The Dynamic Role of Exports in Economic Development: The Fishmeal Industry in Peru, 1956–1966" (unpublished Ph.D. dissertation, Department of Economics, Massachusetts Institute of Technology, 1967), pp. 77–87.

three factor constraints. At different levels of the rate of growth of national income, first foreign exchange, then investment resources, and finally skilled labor might be expected to constrain further growth. The placing of the constraints in the figure is arbitrary. A less developed economy need not necessarily meet the constraints in the order shown. It is also very possible that the constraints could shift so that the economy would be bound by one, then another, and then the first once again.

One way to measure the supply of factors by an export industry to the domestic economy is to use input-output analysis. In brief, given intersectoral flow coefficients, sectoral factor input coefficients, and levels of factor availabilities, one can obtain estimates of allowable annual amounts of sectoral gross output, value added, and deliveries to final demand and at the same time identify amounts of factors used and the binding factor constraint in each year. It is then possible to calculate the shadow prices of the binding factors.

The factors limiting further output in the input-output approach can be cross-checked through use of the binding constraint indicators suggested by Chenery and Strout.[2] The foreign exchange constraint applies if foreign exchange reserves are low or falling, there is excess productive capacity, the marginal capital/output ratio is higher than normal, capital markets are slack, the marginal import ratio is falling, and/or the supply of imports is rationed. On the other hand, the saving constraint is binding if the opposite considerations hold true.

Linkage Effects

Private foreign investment can result in a second type of indirect gain for the economy if the operation of the industry that it creates increases the profitability of other domestic industries through intersectoral relationships or linkages. Linkage effects in this analytical framework are defined as economic benefits or costs resulting from relationships with another industry. It is important to point out that the mere existence of intersectoral flows does not necessarily create a benefit or a cost.

Benefits associated with linkages generally result from the occurrence of one or more of three phenomena—economies of scale, positive externalities, and employment of unemployed or underemployed resources. Linkage costs result from the existence of negative externalities and the creation of unemployment of local factors. Economies of scale arise when an industry, operating on a downward-sloping portion of its average cost curve, experiences additional demand for its output, thereby allowing it to reduce its unit cost of production. Externalities in production are nonappropriated benefits or costs that result when the newly created industry furnishes free inputs to other industries; more precisely, through its activity, the new industry provides a nonpaid factor in another industry's production function.[3] Gains from the use of unemployed or underemployed resources, all of which must be considered net of opportunity costs, occur where institutional barriers have restricted government policy options, thus preventing optimal allocation of resources, or where technological requirements have led to unemployment with full price flexibility.

There would be further linkage benefits if, for economic or psychological reasons, additional amounts of private foreign investment entered the less developed economy following the initial investment. To analyze the overall impact of these further rounds of private foreign investment one can utilize the analytical framework outlined in this chapter for the initial foreign investment, if in so doing, possible double-counting of benefits is avoided.

For purposes of convenient analysis, linkages can be divided into two categories—immediate and fiscal—each with three identical subdivisions. Immediate linkages directly involve operations of the industry created by the foreign investment, whereas fiscal linkages comprise the intersectoral effects associated with government expenditure of incremental revenues collected from expansion of the given industry. The benefits and costs derived within these two categories may support or offset one another.

Within each of these categories, it is useful to differentiate three specific types of linkage effects—investment (including backward

and forward), final demand, and technological.[4] With all three types, welfare gains or losses result from the occurrence of one or more of the three sources of gain or loss listed above. Investment linkages require that the newly expanded industry use inputs supplied by backward-linked domestic industries, or that its outputs be used as intermediate inputs by forward-linked local industries. Final demand linkages refer to any additional stimulus to local industry that might result from the expenditure of factor incomes associated with the expanded industry, as measured above under the direct contribution. Technological linkages comprise benefits or costs that operation of the newly expanded industry or sector imparts to other industries not directly tied through intersectoral flows or expenditures of factor payments. Measurement of these effects requires detailed knowledge of intersectoral flows, industrial production processes, and consumer behavior and thus often must be approximated.

Backward linkages concern inputs demanded by the new industry. Industries may be established to supply inputs into the new industry. Furthermore, certain industries already established (e.g. electricity, transport) can expand output to meet the demand of the new industry and might thereby enjoy (additional) economies of scale. This type of linkage derives from the provision of intermediate inputs of goods and services as well as from sales of capital goods to the industry. To cite just one example, if the output of a backward-linked industry cannot be traded internationally and if economies of scale are achieved by supplying the new industry, then the economy receives an indirect linkage benefit from the private foreign investment. It is important to note Baldwin's provisos concerning the likely prevalence of backward linkages.[5] The first is that backward linkages are unlikely if increasing returns are so significant that supplying industries are able to overcome transportation costs and locate abroad. Second, if certain domestic factor inputs are sufficiently scarce so that inputs must be imported, then the cost of establishing backward-linked industries might become prohibitive.

Forward linkages arise from the sale of outputs of the new industry for use as inputs in other industries. Unlike backward-linked industries that have a captive demand from the new industry, forward-linked industries require sufficient domestic or external demand before they can be established.

A first approximation of an economy's capacity for investment linkages can be attained through input-output analysis.[6] To the extent that low or zero coefficients exist in an input-output matrix, the opportunities for potential investment linkages that existed during the year for which the matrix was calculated are correspondingly small. An input-output matrix, however, is not necessarily a good indicator of possible future linkages, since these would involve changes in the matrix itself. As indicated above, the ultimate reckoning of linkage benefits and costs must consider the opportunity costs of factors and material inputs and the alternative marketability of outputs.

Final demand linkages refer to any additional stimulus to local industries that might result from the expenditure of incremental factor incomes associated with the new industry's factor payments. Naturally the magnitude of this stimulus is diminished by the extent of factor opportunity costs. This type of linkage is analogous to investment linkages but has different timing. Final demand benefits arise from one or more of the three phenomena mentioned above—economies of scale, externalities, and employment of underused resources.

In the first instance, final demand linkages concern that portion of the local consumer market developed by the newly created industry. Interest is focused on the proportions of the associated incremental income that are saved, spent on imported goods, or spent on domestic goods.[7] The concept of final demand linkages implies that factors would not receive equivalent marginal value products in alternative employment. This assumption could very well be valid for an economy with perfectly functioning markets. The existence of imperfect markets in most underdeveloped economies, evidenced by factor immobility, price rigidity, isolated and

Foreign Investment in a Less Developed Economy 49

fragmented markets, and ignorance of technological possibilities, serves to underscore its reasonableness.[8]

The means of estimating final demand linkages are straightforward in theory but difficult in practice. First of all, it is necessary to find the amount of factor payments made by the new industry. These are then compared with each factor's likely income from alternative employment, and the net incremental factor income is calculated. This net incremental factor income must equal the net gain calculated in connection with the new industry's direct contribution to national income. Next, one must estimate the expenditure patterns of those who receive factor payments through knowledge of propensities to import and save compared with the propensity to purchase local goods and services. Finally, it must be determined whether the additional final demand created as a result of the foreign investment will provide sufficient stimulus to aid the new establishment or expansion of local industry, the benefits of which must be considered net of opportunity costs.

Technological linkages comprise benefits or costs that operation of the newly created industry imparts to other industries not directly tied through intersectoral flows or expenditures of factor payments. This type of linkage involves externalities emanating from the new industry to all parts of the domestic economy that are not otherwise directly related. Examples would include labor and management training by the new industry of personnel subsequently employed outside the industry, the industry's direct or indirect promotion of mass education, the spread of new technologies or methods of organization initially brought into operation by the industry, and the industry's construction of infrastructure that is used at less than marginal cost by other parts of the economy.

Fiscal linkages, as earlier defined, are external effects associated with the expenditure of the new industry's payments to government. The underlying assumption is that these amounts, though not usually earmarked for specific uses, when spent could involve linkage effects wholly analogous to those emanating from the newly created industry itself. A government of a low income nation can

thus use fiscal linkages to extend or partially offset any linkage effects stemming from the newly created industry. A special type of fiscal linkage is the tremendous budgetary flexibility that the government gains when large revenues are associated with the industry's operation. This often offers additional freedom in making important economic policy choices. For example, the country may decide to utilize extra government revenues associated with an expanding new industry to substitute for revenues formerly derived from other sources or, alternatively, to provide direct subsidies in lieu of less efficient tariff protection.

A Requirements Analysis Employing Input-Output Techniques

A requirements analysis based on an input-output model allows estimation of total value added and determination of binding factor constraints and shadow prices.[9] Identification of the binding factor not only sheds light on the indirect benefits derived from factor contributions but also assists in evaluating the direct contribution and linkage effects by clarifying the problem of measuring opportunity costs.

It is helpful to review the general approach and major limitations of input-output analysis as it is applied in this context.[10] First, a table of flows differentiates the new industry from the remainder of the economy. Then the matrix of production coefficients and the Leontief inverse matrix are calculated to measure both direct and indirect effects of expanding production by the new industry. Ideally input-output coefficients should be derived from a large sample of firms in which each firm's inputs are regressed on its output, but more often the coefficients are found through the use of aggregate data. The major limitations of input-output analysis are the required assumptions of fixed coefficients and the absence of technological change, though sensitivity analysis can be used to test the restrictiveness of these arbitrary assumptions.[11] In addition, the treatment of investment generally raises difficulties, and there are usually very serious problems raised by the quality and interpretation of data used in the construction of input-output tables in less developed countries.

The input-output model employed in a requirements analysis can be described compactly with the aid of matrix notation.[12] The basic equation of the model is

$$T = F \cdot (I - A)^{-1} \cdot Y$$

where

$T \equiv m \times 1$ vector of total factor use;

$F \equiv m \times n$ matrix of factor requirement coefficients, f_{kj}, of factor k input required to obtain one unit of sector j output;

$I \equiv n \times n$ identity matrix;

$A \equiv n \times n$ matrix of Leontief input-output coefficients, a_{ij}, of sector i input required to obtain one unit of sector j output; and

$Y \equiv n \times 1$ vector of n sectoral levels of deliveries to final demand.[13]

It might be helpful to describe the basic equation verbally. In a given year the level of employment of each factor entering the analysis can be found by multiplying the matrix of factor requirement coefficients (F) by the matrix of final plus intersectoral requirement coefficients $[(I - A)^{-1}$ or the Leontief inverse matrix] and then multiplying this product by the vector comprising sectoral deliveries to final demand (Y). In other words, in order to achieve the level of final demand arrived at by summing all of the elements of vector Y, the economy must employ the amounts of factors individually represented as elements of vector T. The model thus gives factor requirements associated with sectoral levels of final demand in light of economically and technologically determined factor requirement and intersectoral requirement matrices. One additional piece of information is needed before the model can be employed to identify binding factors. This is the vector for each year of factor availabilities, termed U with dimensions $m \times 1$.*

It should be noted that data requirements for the model are large. An input-output matrix, disaggregated sufficiently to allow concentration on the sector or subsector in which the private for-

* In general, of course, if A and F are known, one cannot specify both T and Y. The relationship between T and U is made apparent below.

eign investment occurs, must be available. The impact of the newly created industry is best seen if it is possible to segregate it as a sector. In addition, one must know enough about each entering sector's factor proportions to be able to construct a reasonable approximation of the factor requirements matrix. For some historical years, the T vector of total factor use and the Y vector of final demand by sector are known. It is then possible to cross-check any given year's estimated F matrix by plugging these known values into the basic equation. This can be done for all historical years for which data are available to obtain a representative F matrix for projection purposes.

To employ the model, one first chooses a base year from which the projection is to take place. In the usual instance, the base year Y and T vectors and Leontief inverse matrix are known. The F matrix is then estimated on the basis of historical experience, and its feasibility is checked by using the basic equation. Unless one possesses special information about changing sectoral requirements, the Leontief matrix and the F matrix are assumed to remain fixed over the period of the projection. In general, it is necessary to postulate a target set of sectoral growth rates for Y on the basis of historical experience or special knowledge.* The T vector of factor use for the first year of the projection is found by using the basic equation. This is compared with the separately projected first year U vector of factor availabilities. If all elements of U exceed the corresponding elements of T, the maximum feasible size of the Y vector is found by increasing Y continually until one or more elements of T equal their counterparts in U (all others, of course, still being less). Conversely, if one or more elements of T are initially greater than associated elements of U, target Y is decreased until T is less than or equal to U throughout. In the process of finding maximal Y subject to the factor availability constraints, the factor (or factors) that binds further output is determined. It is

* Since sectoral deliveries to final demand, gross outputs, and value added are all directly related in an input-output system, it is possible to postulate sectoral growth rates for any one of the three to employ the projection model. See the related discussion on pp. 120–23.

then possible to measure the importance of an additional unit of the binding factor, i.e. that factor's shadow price.

Requirements for the second year of the projection are found in exactly the same way as those for the first. In this model the second year can be related to the first by making an assumption concerning investment resources. Recognizing the gestation period inherent in investment, it is fairly crude but reasonable to postulate that some given percentage of output in any one year will translate into investment in the following year. Each year of the projection can then be treated in turn with successive annual identifications of the binding factor constraint, determinations of sectoral output levels, and estimations of shadow prices. Finally, sensitivity analysis for all years can be undertaken to test the importance of altering assumptions regarding any or all of the projected variables, including, for example, the fixed coefficients postulated in the F and $(I - A)^{-1}$ matrices.

In general, more details are known regarding the sector that receives the private foreign investment. If so, then it makes sense to construct separate projected annual entries in the A and F matrices and in the Y vector for this sector. Moreover, the contributions of the sector in question to the annual U vectors of factor availabilities can be segregated so as to gauge their relative importance. In fact, this model is ideally suited for instances in which the flow of private foreign investment results in large and growing contributions of one or more potentially scarce factors.

This general approach also provides the first step in estimating some future linkage effects. Flows into the sector of interest are a necessary but certainly not sufficient condition for the existence of intersectoral backward-linkage effects.[14] The limit of these backward-linkage effects is indicated by the given sector's gross output, which can then be compared with its value added to discover the annual amounts of inputs purchased by the sector in question from the remainder of the economy. All economic benefits that might accrue as a result of these intersectoral flows must be measured net of opportunity costs and net of any negative impacts that occur. The model also helps to determine the importance of final

demand linkages by identifying the binding factor in each year of the projection, thereby aiding the estimation of factor opportunity costs.

In summary, use of the requirements analysis helps determine the projected impacts of direct private foreign investment in several respects. First, it facilitates measurement of the net direct contribution by identifying which factor is binding in each year of the projection and assisting the determination of factor opportunity costs. Second, it permits evaluation of factor contributions associated with the private foreign investment by pinpointing constraining factors in each year and allowing calculation of accompanying shadow prices. Finally, it offers limits regarding the likely importance of economic benefits deriving from backward linkages.

CHAPTER FIVE

The Direct Contribution of Petroleum to Income

This chapter takes the first step in gauging the impact of petroleum on the Nigerian economy. Following the estimation here of the direct contribution of oil to Nigerian income, Chapters Six and Seven deal with the two types of indirect contributions outlined in the preceding chapter. As the theoretical discussion of Chapter Four emphasized, the direct gain from foreign investment is best measured by the total incremental factor payments associated with the foreign investor's expenditures, i.e. the total factor payments net of social opportunity costs. For comparative purposes, the analysis below begins with a standard calculation of the gross contribution of the petroleum industry to Nigerian value added. Focus then shifts to the net concept in an attempt to estimate the net gain and the net gain coefficient for recent years under various assumptions.

The Gross Direct Contribution

A logical way of beginning the calculation of gross value added of the petroleum industry is to review production and export performance. Table 5.1 displays production and export data covering the eleven-year history of petroleum exports from Nigeria. Proceeds from export sales constitute nearly all of the Nigerian petroleum industry's earnings, with the remainder generated mostly by local sales of crude for processing in Nigeria's refinery and by local sales of natural gas. In mid-1967 the rapid growth of the industry's output and hence its contributions to the Nigerian economy were interrupted by civil war disturbances that caused the cutbacks in oil production evidenced in Table 5.1.

TABLE 5.1. CRUDE PETROLEUM PRODUCTION AND EXPORTS, 1958–68

Year	Volume (thousand barrels per day)		Value (million £ Nigerian)	
	Production[a]	Exports	Production	Exports
1958	5	5	0.9	0.9
1959	11	11	2.6	2.6
1960	17	17	4.2	4.2
1961	46	46	11.3	11.3
1962	68	68	17.2	17.2
1963	76	76	20.1	20.1
1964	120	120	32.0	32.0
1965	270	266	69.1	68.1
1966	415	383	99.7	92.0
1967[b]	317	300	76.6	72.4
1968[b]	142	142	36.6	36.6

Data for 1958–65 from Nigeria [6], p. 11; unpublished individual company data for 1966–68. Production exceeds exports beginning in late 1965, when Nigeria's refinery started operating; the refinery was officially shut down in July 1967, owing to the civil war.

[a] For the years 1958–64: all Shell-BP; 1965: Shell-BP, 239, Gulf, 27; 1966: Shell-BP, 320, Gulf, 51, Safrap, 12; 1967: Shell-BP, 226, Gulf, 55, Safrap, 19; 1968: Shell-BP, 44, Gulf, 98.

[b] The Nigerian civil war caused Shell-BP and Safrap to discontinue production in July 1967; Gulf continued producing and exporting throughout 1967 and 1968, and Shell-BP resumed production in October 1968.

In the first instance the magnitude of the petroleum industry's gross value added depends on the value of crude oil exports and local sales. To find value added as a residual, various deductions can be made from the industry's total proceeds. Alternatively, value added can be built up from individual components. Both of these methods are described in Table 5.2, which contains calculations of the Nigerian oil industry's value added for the six years during which it was at all significant.*

For an underdeveloped economy with an important foreign-owned export sector, Gross National Product is generally a more useful national accounting measure than Gross Domestic Product. GDP exceeds GNP by the amount of net factor payments made abroad. These payments are often very large where foreign export industries are involved. GDP accounting thus gives an inflated view of the amount of goods and services actually available to indigenous owners of factors of production. This bias is shown in Table 5.3, where contributions of the Nigerian petroleum industry to

* For an algebraic discussion of these concepts, see Appendix E.

TABLE 5.2. PETROLEUM INDUSTRY VALUE ADDED, 1963–68
(Million £ Nigerian, except as otherwise indicated)

Value	1963	1964	1965	1966	1967	1968
VALUE ADDED TO GROSS DOMESTIC PRODUCT						
Calculated as a residual:						
1. Proceeds from export sales	20.1	32.0	68.1	92.0	72.4	36.6
2. Proceeds from local sales	0.6	0.7	1.0	8.6	8.2	0.3
3. Net foreign capital flow	5.0	18.0	17.5	28.8	43.5	40.7
4. Imports of materials	4.2	11.7	13.5	19.5	13.7	13.6
5. Imports of services	6.4	12.2	22.2	38.5	37.5	34.8
6. Local goods and services	7.2	10.3	18.5	27.9	21.9	11.2
7. Value added to GDP^a (1 + 2 + 3 − 4 − 5 − 6)	7.9	16.5	32.4	43.5	51.0	18.0
Calculated from individual components:						
8. Payments of government	5.0	12.3	13.3	18.7	27.0	15.8
9. Harbor dues and port charges	0.6	1.0	2.0	2.7	2.4	0.4
10. Local wages and salaries	2.0	2.3	2.7	2.7	2.5	1.8
11. Net factor income paid abroad^c	0.2	1.0	14.2	19.4	19.0	—^{b}
12. Value added to GDP^a (8 + 9 + 10 + 11)	7.9	16.5	32.4	43.5	51.0	18.0
VALUE ADDED TO GROSS NATIONAL PRODUCT						
Calculated as a residual:						
13. Value added to GNP^a (1 + 2 + 3 − 4 − 5 − 6 − 11)	7.7	15.5	18.2	24.1	32.0	18.0
Calculated from individual components:						
14. Value added to GNP^a (8 + 9 + 10)	7.7	15.5	18.2	24.1	32.0	18.0

Data from unpublished individual company materials. The calculations were made before rounding and hence figures shown above do not always sum exactly.
^{a} May not equal sum of components because of rounding.
^{b} — indicates less than £500,000 Nigerian.
^{c} This entry includes depreciation, dividends, profits, interest, and all other items remitted abroad.

GDP and to GNP are compared. In 1966, the last full year of normal oil industry operation, petroleum exploration and production contributed 1.7 per cent of total Nigerian GNP, significantly less than its 3.0 per cent share of GDP in that year.

It would be better to calculate petroleum industry value added to GDP on the basis of realized rather than posted prices. This would require data that are not available, i.e. information on each producing company's average annual realizations. The degree of

TABLE 5.3. THE CONTRIBUTION OF PETROLEUM VALUE ADDED TO NIGERIAN GDP AND TO NIGERIAN GNP, 1963–68
(Million £ Nigerian, except as otherwise indicated)

Value	1963	1964	1965	1966	1967	1968
Gross Domestic Product (GDP) in market prices:						
Petroleum value added	8	17	32	44	51	18
Nonpetroleum value added	1,303	1,308	1,381	1,419	1,434	1,463
Gross Domestic Product	1,311	1,325	1,413	1,463	1,485	1,481
Petroleum value added as per cent of GDP	0.6	1.3	2.3	3.0	3.4	1.2
Gross National Product (GNP) in market prices:						
Petroleum value added	8	16	18	24	32	18
Nonpetroleum value added	1,286	1,284	1,366	1,401	1,413	1,443
Gross National Product	1,294	1,300	1,384	1,425	1,445	1,461
Petroleum value added as per cent of GNP	0.6	1.2	1.3	1.7	2.2	1.2

Data from Table 5.2 and Appendix Table A.1.

error involved in using posted prices (including discounts) probably amounts to an overstatement of petroleum value added to GDP of about 20–25 per cent in 1966 (the first year affected) and in 1967.* If the more appropriate GNP accounting is employed, it makes no difference whether realization or posted prices are used to evaluate output of petroleum, since net factor payments made abroad are excluded from GNP. (See Appendix E.)

The Net Direct Contribution

The task of this section is to find the net direct contribution of petroleum to Nigerian income. After presentation of the relevant data, four separate solutions for the net gain and the net gain coefficient are found employing combinations of two sets of assumptions. For 1963–68 the direct impact of the exploring and producing companies, hereafter referred to as the petroleum industry, is investigated under two different assumptions regarding social opportunity costs. And for 1965 only, the definition of the petroleum

* In 1968, if net factor income paid abroad on petroleum account was negligible, as reported, then no error is introduced, since the industry's contribution to GDP equaled its contribution to GNP. See Table 5.2 and Appendix E.

The Direct Contribution of Petroleum to Income 59

industry is expanded to take into account the local and international firms that service or supply the exploring and producing companies, hereafter termed the ancillary firms, again using alternative assumptions for opportunity costs.

Total receipts of the petroleum industry. Total receipts of the petroleum industry are defined as comprising the total proceeds from export sales plus proceeds from local sales received by the aggregate of the companies exploring for and producing oil in Nigeria. It will be recalled that the annual net direct contributions of the petroleum industry are compared with annual total receipts to calculate annual net gain coefficient. For the six years in question, 1963–68, the yearly amounts of total receipts of the petroleum industry in Nigeria were (in millions of £ Nigerian): 1963, 20.7; 1964, 32.7; 1965, 69.1; 1966, 100.6; 1967, 80.6; and 1968, 36.9.*

Factor payments and factor shares. Knowledge of the split of the petroleum industry's annual expenditures among factors and materials is essential for the calculation of the industry's net direct contribution.† To arrive at factor shares of investment in petroleum, one can put together a Leontief production function, i.e. the petroleum sector's column in a Leontief input-output matrix. Each element of this column is simply a factor share found by taking the ratio of the factor payment to total expenditure.

Table 5.4 presents a summary Leontief function for the production of crude oil in Nigeria, using 1965 data. Two difficulties are posed by this formulation. The first arises from the fact that choosing a single year to represent what might be considered normal production activity is a very shaky procedure unless due qualifications are made. Petroleum production is a long-cycled process in which the mix of inputs as measured by company expenditures shifts quite markedly over time. Results for a country like Nigeria, where

---

* These figures are aggregations of individual company unpublished data for proceeds from export sales plus proceeds from local sales, as reported in the first two lines of Table 5.2.

† In making the net gain calculation, intermediate materials can for expositional purposes be given shares of expenditure without affecting the results, since expenditures on materials never enter into value added.

TABLE 5.4. LEONTIEF PRODUCTION FUNCTION FOR THE PETROLEUM INDUSTRY, 1965 DATA

Industry expenditure categories	Industry expenditures (million £ Nigerian)	Input coefficients, based on	
		Net total expenditures in Nigerian currency	Total expenditures
Expenditures in Nigerian currency:			
Local purchases of goods (including fuel and lubricants)	3.903	0.117	0.057
Well-site preparation and roads	3.086	0.093	0.045
Drilling	3.000	0.090	0.043
Transportation	2.474	0.074	0.036
Construction	1.395	0.042	0.020
Seismic surveys	1.183	0.036	0.017
Dredging	0.522	0.016	0.008
Catering	0.230	0.007	0.003
Other local expenditures	2.711	0.081	0.039
Wages and salaries	2.706	0.081	0.039
Payments to the Nigerian government	13.345	0.401	0.193
Harbor dues and port charges	1.997	0.060	0.029
GROSS TOTAL	36.552	—	—
Residual (interest, depreciation, profits, etc.[a]	−3.233	−0.097	−0.047
NET TOTAL	33.319	1.001[b]	—
Imports of goods and services	35.733	—	0.517
Total proceeds equal total expenditures	69.052	—	0.999[b]

Aggregation of individual company unpublished data; see Appendix C.
[a] Residual is calculated as total proceeds (i.e. exports equal 68.097 plus local sales equal 0.955) of 69.052 less imports of goods (13.518) and services (22.215) of 35.733 less gross total expenditures in Nigerian currency (36.552).
[b] Totals may not add to 1.000 because of rounding.

the industry for the most part is young and expanding, might be especially unreliable. Second, it is not possible to gain an entirely accurate idea of factor intensities from Table 5.4, because the Nigerian petroleum industry subcontracts a large portion of its production activity to ancillary firms. (For a detailed examination, see Chapter Seven and Appendix C.) This means that the actual purchase of materials, labor, capital, and other inputs is postponed a step.

Table 5.5 is the result of an effort to mitigate, at least partially, the first of these two shortcomings. In this table, factor and mate-

The Direct Contribution of Petroleum to Income 61

rial payments and shares pertaining to the exploring and producing companies are presented for all six years of interest. In light of the comparative youth of Nigeria's petroleum industry and of the disruptions it has recently faced, it is not easy to discern the beginnings of any meaningful trends from this table. The following likely trends might be kept in mind, however, in interpreting

TABLE 5.5. COMPARISON OF FACTOR SHARES FOR THE PETROLEUM INDUSTRY, 1963–68

Industry expenditure categories	1963	1964	1965	1966	1967	1968
INDUSTRY EXPENDITURES (millions of £ Nigerian)						
Local purchases of goods and services	7.2	10.3	18.5	27.9	21.9	11.2
Local wages and salaries	2.0	2.3	2.7	2.7	2.5	1.8
Payments to the Nigerian government	5.0	12.3	13.3	18.7	27.0	15.8
Harbor dues and port charges	0.6	1.0	2.0	2.7	2.4	0.4
Gross total expenditures in Nigerian currency[a]	14.8	25.9	36.6	52.0	53.8	29.2
Residual (interest, depreciation, profits, etc.)[b]	−4.7	−17.0	−3.3	−9.5	−24.5	−40.7
Imports of goods and services	10.6	24.0	35.7	58.0	51.2	48.4
Total proceeds equal total expenditures	20.7	32.8	69.1	100.6	80.6	36.9
FACTOR SHARES BASED ON TOTAL EXPENDITURES[c]						
Local purchases of goods and services	0.346	0.315	0.268	0.277	0.277	0.304
Local wages and salaries	0.097	0.070	0.039	0.027	0.031	0.051
Payments to the Nigerian government	0.242	0.375	0.193	0.186	0.335	0.428
Harbor dues and port charges	0.030	0.030	0.029	0.026	0.030	0.011
Gross total expenditures in Nigerian currency	—	—	—	—	—	—
Residual (interest, depreciation, profits, etc.)[b]	−0.228	−0.520	−0.047	−0.094	−0.304	−1.103
Imports of goods and services	0.513	0.731	0.517	0.577	0.636	1.312
Total proceeds equal total expenditures	1.000	1.000	0.999	0.999	1.000	1.001

Aggregation of individual company unpublished data.
[a] May not equal sum of components because of rounding.
[b] Residual is calculated as total proceeds less imports of goods and services less gross total expenditures in Nigerian currency.
[c] All calculations done prior to rounding.

62   The Direct Contribution of Petroleum to Income

TABLE 5.6. INPUT COEFFICIENTS, 1965 DATA

Industry expenditure categories	Total	Materials		Wages and salaries	Depreciation	Nigerian taxes	Residual[a]
		Local	Imported				
Local purchase of goods (including fuel and lubricants)	0.057	0.003	0.031	0.007	0.004	0.009	0.003
Well-site preparation and roads	0.045	0.004	0.013	0.012	0.003	—	0.013
Drilling	0.043	0.001	0.004	0.008	0.008	0.002	0.020
Transportation	0.036	0.003	0.007	0.007	0.004	—	0.015
Construction	0.020	0.003	0.004	0.006	0.003	—	0.004
Seismic surveys	0.017	0.001	0.002	0.006	0.001	—	0.007
Dredging	0.008	—	0.002	0.004	0.002	—	—
Catering	0.003	0.001	0.001	0.001	—	—	—
Other local expenditures[b]	0.039	0.003	0.008	0.009	0.005	0.002	0.012
Total	0.268	0.019	0.072	0.060	0.030	0.013	0.074

Total input coefficients, based on total expenditures from Table 5.4, column 3; broken down in accordance with Appendix C. Dash indicates that coefficients are less than 0.0005.

[a] Interest, profit, etc.
[b] Disaggregated using average coefficients derived from total of contracting expenditure categories (i.e. "well-site preparation and roads" through "catering" in this table).

the results of this chapter. Local purchases and imports of goods and services as well as wages and salaries should fall proportionately over the next several years. Payments to the Nigerian government and harbor dues and port charges are expected to rise rapidly, both absolutely and as portions of the total.* And the residual category should turn positive and become an increasingly large absolute and proportionate share of total expenditures.

In order to circumvent the second difficulty associated with Table 5.4, it is convenient to consider the firms that supply and service the petroleum industry as if they were in fact a part of this industry. This change of definitional assumptions offers the opportunity for an alternative calculation of net gain. But because of data limitations, this calculation can only be done for 1965, the year for which detailed survey information on the ancillary firms is available.† In Table 5.6 the input coefficient for each of the sup-

* As mentioned in Chapter Two, one should conceive of most if not all of the payments to the Nigerian government as economic rent rather than profits taxes in the usual sense.
† The methodology and results of this survey are contained in Appendix C.

plying or contracting categories is broken down among the six types of inputs indicated. These intermediate results are then added to the input coefficients for the petroleum industry in Table 5.7 to arrive at the consolidated production function of the newly defined industry.

The total input coefficients for the consolidated Nigerian petroleum industry contain no major surprises. As Table 5.7 demonstrates, nearly three-fifths of the total Nigerian petroleum industry expenditure in 1965 consisted of imported goods or services, and almost one-fourth amounted to payments to the Nigerian government or to quasi-governmental agencies. The remaining shares were spread among wages and salaries (10 per cent), purchases of local materials (2 per cent), and residual items (6 per cent). It is of interest to digress briefly in order to compare these results first with other mineral export industries and then with Nigerian manufacturing.

Table 5.8 displays comparative input coefficients for Nigerian petroleum and four other foreign-owned extractive export industries—Venezuelan petroleum, Zambian and Chilean copper, and

TABLE 5.7. LEONTIEF PRODUCTION FUNCTION FOR THE CONSOLIDATED NIGERIAN PETROLEUM INDUSTRY, 1965 DATA

Category	Input Coefficients			Expenditures (thousands £ Nigerian)
	Petroleum Industry	Local suppliers and contractors	Consolidated totals	
Local materials	—	0.019	0.019	1,312
Imported materials	0.196	0.072	0.268	18,506
Imported services	0.321	—	0.321	22,166
Local wages and salaries	0.039	0.060	0.099	6,836
Payments to Nigerian government (including harbor dues and port charges)	0.222	0.013	0.235	16,227
Residual (depreciation, interest, profits, foreign taxes, etc.)	−0.047	0.104	0.057	3,936
Total	0.731	0.268	0.999	68,983[a]

Data from Tables 5.4 and 5.6.
[a] £68,983 is 99.9 per cent of the actual total, £69,052.

TABLE 5.8. COMPARISON OF EXTRACTIVE EXPORT INDUSTRY INPUT COEFFICIENTS FOR SELECTED INDUSTRIES AND COUNTRIES

Category	Petroleum		Copper		Bauxite
	Nigeria 1965	Venezuela 1963–65	Zambia 1956–57	Chile 1956	Jamaica 1961
VALUE (millions of £ Nigerian)					
Total export industry expenditure[a]	69.1	795.0	136.6	118.5	13.1
Imported goods and services	40.7	323.0	62.0[b]	26.3[b]	6.2
Payments to host government	16.2	349.0	32.1	48.1	3.2
Wages and salaries paid locally	6.8	64.0	22.0	16.1	1.4
Materials purchased locally[a]	1.3	30.0	2.2	28.0	0.1
Other local payments	3.9[b]	29.0[b]	18.3		2.2[b]
INPUT COEFFICIENTS (share of total export industry expenditures)					
Total export industry expenditure[a]	0.999	1.000	1.000	1.000	1.000
Imported goods and services	0.589	0.406	0.454[b]	0.222[b]	0.475
Payments to host government	0.235	0.439	0.235	0.406	0.243
Wages and salaries paid locally	0.099	0.081	0.161	0.136	0.106
Materials purchased locally[a]	0.019	0.038	0.016	0.236	0.007
Other local payments	0.57[b]	0.36[b]	0.134		0.169[b]

Data are from the following sources: Nigeria—Table 5.7. Venezuela—G. Gabriel, pp. 172, 184–86. Zambia—Baldwin, pp. 38–39. Chile—Reynolds, pp. 363–69. Jamaica—H. D. Huggins, *Aluminium in Changing Communities* (London: Andre Deutsch, 1965), p. 125. Conversions from local currency to £ Nigerian made by author.

[a] Only the Nigerian figures are adjusted for import content of materials purchased locally.
[b] Estimated as a residual.

Jamaican bauxite. In large part the data underlying these coefficients are sufficiently fragile to obviate the drawing of strong conclusions. Nonetheless an interesting though not unexpected pattern emerges from this table. With the exception of Chilean copper, all coefficients for imported goods and services range between .40 and .60. Coefficients for payments to host governments run between .20 and .45. Nigeria is at the high end of the former range and on

the lower edges of the latter. This is probably best explained by the relatively low taxes paid during early stages of oil production. In all instances the coefficients for wages and salaries paid locally and for locally purchased materials are uniformly low. But it must be recalled that these coefficients apply to large and often rapidly growing absolute amounts, so that the industry's impact on a host economy can be significant notwithstanding low coefficients for local inputs.

For further perspective a less complete comparison can be made with aggregate data on Nigerian manufacturing, as derived from the Industrial Surveys of the Federal Office of Statistics (covering firms with ten or more employees).[1] Interestingly enough, during each of the three years 1963 through 1965, Nigerian manufacturing (excluding small-scale) paid slightly less in wages and salaries compared with gross output than did the petroleum industry. For example, in 1965, of a total figure for manufacturing receipts of £218.2 million, £20.5 million was paid out in wages and salaries. This implies a wage coefficient of .094 versus petroleum's .099 for the same year. Unfortunately the Surveys give no information about import content. The imported plus local materials coefficient for manufacturing has run at about .45 recently as compared with .29 for the petroleum industry, but there is no way of knowing the split between imported and locally purchased materials. Using slightly different definitions, recent value added in the petroleum industry has had coefficients near .70, whereas value added in manufacturing has been closer to .40. As a result of data limitations and conceptual differences, no very persuasive conclusions can be drawn from this aggregated comparison of petroleum and manufacturing.

In short, the process of producing petroleum in Nigeria is characterized by the following properties. It employs large capital inputs and requires advanced technological skills. Its material inputs for the most part involve technologically advanced methods of production and have sufficiently high value/transport ratios so that much the largest portion of materials must be imported at least initially. And its labor requirements result in the need for relatively

large amounts of skilled human inputs with virtually no seasonality or normal cause for labor turnover. Though there is some room for substitution, current technology used to produce crude petroleum is such that noneconomic goals would have to take precedence before this picture could be changed much. Hence, the production function for oil in Nigeria is conducive to large economic rents, part of which translate into payments to government, moderate but significant levels of wages and salaries, and small use of local materials and concomitantly large imports of materials and services.

Factor opportunity costs. The elusive concept of factor opportunity costs creates grave difficulties in the calculation of the petroleum industry's net direct contribution. For this reason two different assumptions are employed to allow an approximation of net gain under each of the two definitions of the Nigerian petroleum industry developed in the preceding section. In the first instance, it is postulated that foreign exchange has been the factor constraining increases in Nigerian output during the period under consideration. Considerable evidence exists to back up this assumption. With the assumption of a binding foreign exchange constraint, one can formally though somewhat artificially impute a zero opportunity cost to all other factors. Naturally, the opportunity cost of all materials must be 100 per cent to avoid double-counting in the calculation of value added.

From the end of 1959 to the end of 1968, Nigerian reserves of foreign exchange dropped from £223 million to £47 million, the last-mentioned figure being only about two months' worth of nonpetroleum imports.[2] Most of this decline took place prior to the civil war, but the war seriously exacerbated the already strong pressures on Nigerian foreign exchange reserves. This decline is a good indication that Nigerian growth has recently been constrained by a foreign exchange shortage. Other Chenery-Strout indicators point to this same conclusion. Reference to Appendix Tables A.1 and A.2 shows that Nigeria's marginal import ratio has been falling and negative recently (even if allowances are made for varying geographic coverage). In addition, partly because of

The Direct Contribution of Petroleum to Income 67

the war, Nigeria was forced to adopt an import rationing system in 1968. Finally, excess productive capacity exists in Nigeria's nascent industrial structure, and her little-developed capital market is rather sluggish.

But even if Nigeria has indeed been confronted with a foreign exchange constraint, this does not necessarily imply that skilled and unskilled labor and natural resources have had zero opportunity costs. Hence it is useful to make an alternative assumption regarding the opportunity costs of local factors employed by the petroleum industry. In fact, one could argue that skilled labor would receive nearly its equivalent payment in alternative employment, the possible difference reflecting the inducement necessary to attract the skilled labor to petroleum. The opportunity cost of unskilled labor in Nigeria is surely very low. For the country as a whole, it is certainly not zero. But in face of the large numbers of urban unemployed, it can be taken as virtually equal to zero for the oil industry. In practice, this distinction is not very important, since the changes brought about by varying assumptions concerning the opportunity cost of unskilled labor are very small. Finally, payments to government, reflecting mainly rents associated with the consumed natural resources, should not be assumed to be totally without opportunity costs.

Calculation of the net direct contributions and net gain coefficients. It now remains to put together the data and assumptions set forth in the preceding sections in order to approximate the annual net direct contributions of petroleum and the associated net gain coefficients. Four alternative assumed situations are examined. Consider first the petroleum industry defined to include only the international firms that explore for and produce oil in Nigeria. If a binding foreign exchange constraint is assumed so that in a formal sense all other factors have a zero opportunity cost, then the net annual contributions to Nigerian income equal the sum of payments to government plus harbor dues and port charges plus local wages and salaries. For 1963–68, the net contributions (in millions of £ Nigerian) are 7.6, 15.6, 18.0, 24.1, 31.9, and 18.0. (These items are found as totals of lines 8, 9, and 10 in Table 5.2.)

The net gain coefficients found by comparing the petroleum industry's annual contributions with its total receipts for the years 1963–68, respectively, are .37, .48, .26, .24, .40, and .49. Hence, it can be concluded that between one-fourth and one-half of the petroleum industry's total receipts have recently been translated into factor incomes in Nigeria.

If one takes an alternative view of factor opportunity costs and assumes that some of the complementary Nigerian factors would not be unemployed in the absence of petroleum, this result would be altered. There are no very convincing grounds for assessing opportunity costs of labor and natural resources in Nigeria. For sake of argument, the following arbitrary assumptions are employed regarding percentages of factor payments to be considered as factor opportunity costs: wages and salaries, 75; harbor dues and port charges, 50; and payments to government, 20. The estimated annual net direct contributions and net gain coefficients, respectively, are then reduced to the following (contributions in millions of £ Nigerian): 1963, 4.8, .23; 1964, 10.9, .33; 1965, 12.3, .18; 1966, 17.0, .17; 1967, 23.4, .29; and 1968, 13.2, .36. This changed view of factor opportunity costs thus lowers the net gain to Nigeria from petroleum to a range encompassing one-fifth to one-third of the industry's total receipts.

It has been suggested in the foregoing discussion that the scope of definition of the petroleum industry can usefully be widened to include the ancillary servicing and supplying firms as well as the exploring and producing companies. But because of limitations on the availability of data, analysis of net gain on this basis can only be carried out for the single year 1965. Clearly, the net contribution should be larger with this broader definition. Assuming zero opportunity cost for all factors other than foreign exchange, the 1965 net contribution, found by summing local wages and salaries plus payments to Nigerian government (including harbor dues and port charges) in Table 5.7, becomes £23.1 million. Comparison of this figure with total receipts, redefined to include gross receipts of the ancillary firms from the petroleum industry in 1965, indi-

The Direct Contribution of Petroleum to Income 69

cates that the net gain coefficient would be .25 under this mix of assumptions.*

Finally, if the broader definition of the oil industry is retained and combined with the more restrictive assumptions regarding factor opportunity costs, the outcome is somewhat less optimistic. The net gain for 1965 works out to £14.2 million, whereas the coefficient reduces to .15. Overall, on any reasonable set of assumptions, it is likely that roughly one-fourth of gross output of the petroleum industry in Nigeria has translated into net contributions to Nigerian income. But in light of the fact that payments to government are projected to rise proportionately more rapidly than total receipts, the net gain coefficient of petroleum in Nigeria should increase significantly in the future. (See the related discussion in Chapter Eight.)

* The ancillary firms' gross receipts from the petroleum industry in 1965 were £23.3 million, as reported in Table C.1, Appendix C.

CHAPTER SIX

Factor Contributions to Other Sectors

In addition to the direct gains evaluated in Chapter Five, private foreign investment carried out by the petroleum industry in Nigeria has indirectly benefited the Nigerian economy. This chapter considers one category of indirect benefits, the contribution of factors of production to the nonpetroleum sectors of the Nigerian economy. Chapter Seven will investigate the other category, linkage effects.

Chapter Four contained a theoretical discussion of factor contributions. A combination of institutional, technological, and economic influences dictates the aggregate demands for factors of production in a less developed economy over a short period of time (say one year). Given the underdeveloped nature of such economies, it is highly unlikely that the aggregate supplies of factors precisely match the aggregate demands for them. In the usual situation, most factors are in excess supply and only one or a few factors are scarce. The economy is thus constrained from achieving further additions to output by the limited supply of this binding factor. Additions of this factor would allow further growth to the point where another factor would begin to constrain.

This factor constraint approach is now followed in analyzing the contribution of factors of production by the Nigerian petroleum industry. The petroleum industry's payments to the Nigerian government lead to potential public domestic saving and thus contribute to investment resources. Petroleum export earnings and net private foreign capital inflows provide foreign exchange in amounts significantly in excess of industry requirements on cur-

Factor Contributions to Other Sectors 71

rent account. And the oil industry's initial importation of management and skilled labor, followed by its gradual training of indigenous replacements and future local entrepreneurs, furnish skilled manpower to the Nigerian economy. At a given point in time any or all of these factor contributions might allow additional Nigerian economic development.

Foreign exchange has been the constraining factor of production in Nigeria (see pp. 66–67). The petroleum industry has contributed foreign exchange for use by other sectors of the Nigerian economy, and it has also supplied investment resources and some skilled labor, factors that might have been binding if more foreign exchange had been available. It is thus of interest to consider all three types of factor contributions. For expositional reasons that will become apparent, contributions to investment resources are treated first, followed by foreign exchange and skilled labor.

Investment Resources

Tables 6.1 and 6.2 summarize the available information concerning the various categories of payments that the petroleum industry makes to the Nigerian government.* Excluding 1961 and 1962, when premiums for new concessions swelled the totals, petroleum industry payments did not exceed £5 million in Nigeria until 1964. Accumulated capital allowances have kept the payments to date at fairly modest levels compared with figures for output and gross receipts.†

That oil-derived revenues have not yet begun to dominate total receipts in Nigerian public finance is shown in Table 6.3. But in a period of only eight years petroleum revenues have increased from less than 1 per cent of total government revenues to over 13 per cent of a normalized 1967 figure—and this at a time when total revenues were nearly doubling. The current and probable fu-

---

* These data are spread over two tables because of a discontinuity in reporting categories of the individual companies between 1962 and 1963.
† For an analysis of why this is so, see Chapter Two and Appendix B.

TABLE 6.1. PETROLEUM INDUSTRY PAYMENTS TO THE
NIGERIAN GOVERNMENT, 1959–62
(*Million £ Nigerian*)

Payments	1959	1960	1961	1962
Royalties	0.2	0.3	0.4	1.7
Rentals	—[a]	0.7	0.9	2.0
Premiums	—[a]	—[a]	5.7	4.4
All others (customs and stamp duties, petroleum profits tax, other taxes)	0.1	0.3	0.1	0.4
Total payments to government	0.3	1.3	7.1	8.5

Data from Robinson, pp. 224–25; "Nigeria: Growing Source of World Oil," p. 236; Nigeria [10], p. 122; and unpublished individual company data.
[a] Less than £50,000 Nigerian.

TABLE 6.2. PETROLEUM INDUSTRY PAYMENTS TO THE
NIGERIAN GOVERNMENT, 1963–68
(*Million £ Nigerian*)

Payments	1963	1964	1965	1966	1967	1968
Royalties	1.8	2.2	5.3	8.3	14.9	2.5
Rentals	3.0	3.9	4.4	5.1	5.2	6.8
Petroleum profits tax	—[a]	5.3[b]	1.5	2.6	5.5	3.8
Customs and stamp duties, premiums, and other taxes	0.2	0.9	2.2	2.7	1.4	2.7
Total payments to government	5.0	12.3	13.4	18.7	27.0	15.8

Data from unpublished individual company data.
[a] Less than £50,000 Nigerian.
[b] Includes advance payment of £5 million.

ture importance of oil industry payments to government is clearly established.

There has as yet been no Nigerian legislation calling for separate treatment and use of petroleum-derived revenues. In many less developed nations, governments have required that a certain portion of revenues received from a major extractive export industry be placed in a separate development fund. This is certainly a reasonable procedure if it is felt desirable to direct public opinion pressures toward development expenditures, but it begs the economic question of fungibility among government assets. On

Factor Contributions to Other Sectors

TABLE 6.3. COMPARISON OF PETROLEUM AND NONPETROLEUM
NIGERIAN GOVERNMENT REVENUES, 1959–68
(*Million £ Nigerian, except as otherwise indicated*)

Year	Total government current revenues[a]	Petroleum revenues[b]	Nonpetroleum revenues[a]	Petroleum revenues as percentage of total revenues
1959	114.8	0.3	114.5	—[c]
1960	126.6	1.3	125.3	1.0
1961	135.1	7.1	128.0	5.3
1962	140.6	8.5	132.1	6.0
1963	146.8	5.0	141.8	3.4
1964	178.5	12.3	166.2	6.9
1965	190.0	13.3	176.7	7.0
1966	197.1	18.7	178.4	9.5
1967	163.0[d]	27.0	136.0[d]	16.6
1968	155.8[d]	15.8	140.0[d]	10.1

Data from Tables 6.1 and 6.2, Nigeria [12], pp. 107–8; and unpublished estimates for 1967 and 1968.
[a] Fiscal year (starting April 1).
[b] Calendar year.
[c] Less than .05 per cent.
[d] Preliminary estimate, excluding the former Eastern Region.

the other hand, government decision-makers might very well consider incremental revenues (from an export industry in this instance) differently from some anticipated average level of government revenue.

The sum of petroleum-related and all other current revenues must exceed total government current expenditure during a given year if there is to be any public domestic saving. Public saving will increase to the extent that petroleum revenues grow faster than the rate of growth of current expenditures, i.e. there is no substitution of oil for non-oil revenues. In this way the petroleum industry can contribute to domestic saving and hence to total investment resources, a potentially scarce factor of production.*
Table 6.4 documents the mixed public saving achievements of the Nigerian governments. Current revenues including those derived from oil grew at over 8 per cent per annum between 1959 and 1966, but current expenditures mushroomed at an annual rate

* Expansion of the petroleum industry can also contribute to Nigerian investment resources to the extent that local factor payments made by the industry are saved rather than spent on local or imported goods and services. See the related discussion of final demand linkages in Chapter Seven.

TABLE 6.4. NIGERIAN GOVERNMENT'S PUBLIC SAVING, 1959–66
(Million £ Nigerian, except as otherwise indicated)

Year[a]	Total government current revenues	Total government current expenditures	Public saving	Public saving as a percentage of total government current revenues
1959	114.8	94.8	20.0	17.4
1960	126.6	104.1	22.5	17.8
1961	135.1	125.6	9.5	7.0
1962	140.6	131.9	8.7	6.2
1963	146.8	138.5	8.3	5.7
1964	178.5	157.7	20.8	11.7
1965	190.0	187.2	3.8	1.9
1966	197.1	195.6	1.5	0.8

Data from Table 6.3 and from Nigeria [12], pp. 107–8.
[a] Fiscal year (starting April 1).

approaching 11 per cent. Since 1961, in all years save one, everything else being equal, public saving would have been close to zero or negative in the absence of petroleum revenues.

In the process of financing and carrying out their own operations, the petroleum industry and the ancillary firms have naturally increased total fixed capital formation in Nigeria. By the end of 1968 cumulative capital spending by the industry had exceeded £300 million, and that of the suppliers and contractors was probably in excess of £20 million. This information, though of interest in its own right, has no bearing on the analysis of factor contributions to the nonpetroleum sectors of the Nigerian economy. Further discussion of the oil industry's own capital investment is therefore unnecessary in this context.

Foreign Exchange

One of the major vehicles through which an industry established with foreign investment can impart the potential for growth to other sectors in the host economy is the contribution of foreign exchange. In an analysis of the balance-of-payments impact of a large extractive export industry such as petroleum, however, export values alone have relatively little meaning.[1] What really matters in the context of Nigerian oil are the amounts of foreign exchange that the petroleum industry brings into the country to ex-

Factor Contributions to Other Sectors

change for domestic currency in order to make payments to the government or to purchase local goods and services.* In addition, the total direct and indirect import cost of the operations of the petroleum industry is of considerable interest. To the extent that a portion of the capital inflow resulting from investment by the petroleum industry flows back out of the Nigerian economy in the form of expenditures on imports, the scope for direct and indirect benefits accruing to Nigeria is diminished.

Balance-of-payments impacts. There are two alternative methods of arriving at the balance-of-payments impact of the Nigerian petroleum industry—one dealing with domestic expenditures and earnings and the other with international financial flows.† The first is to sum payments to government and all other local payments, including those for local goods and services, local wages and salaries, and harbor dues and port charges, i.e. transactions that cause the industry to purchase and use Nigerian currency, and to subtract proceeds from local sales from this total. The second is to sum export proceeds and foreign capital inflows, including reimbursement of harbor dues paid in Nigeria on behalf of the shipper, and to subtract all foreign capital outflows, i.e. imports of goods and services, net factor income paid abroad, and increases in overseas cash balances, from this total.

In an accounting sense both methods are equivalent. Industry receipts, including export proceeds plus proceeds from local sales plus foreign capital flows, must equal industry expenditures, comprising imports of goods and services plus net factor income paid abroad plus increases in overseas cash balances plus payments to government plus other local payments. The equivalence of the two methods of measuring the balance-of-payments impact can be shown by subtracting imports of goods and services, net factor

---

* In actual practice the industry makes its payments for rents, royalties, and petroleum profits taxes into a Nigerian government account in the United Kingdom in British pounds. But the intent of the statement is still correct, since the Nigerian government at some point exchanges the British pounds, and hence the impact on the economy ultimately is as described.

† Import-substituting effects of the Nigerian refinery are discussed in Chapter Seven, pp. 92–95. For an algebraic discussion of the balance-of-payments impact, see Appendix E.

TABLE 6.5. IMPACT OF THE PETROLEUM INDUSTRY
BALANCE OF PAYMENTS, 1963–68
(*Million £ Nigerian*)

Payments	1963	1964	1965	1966	1967	1968
A. Local currency expenditures:						
+ Payments to government	5.0	12.3	13.3	18.7	27.0	15.8
+ Other local payments	9.8	13.6	23.2	33.3	26.8	13.4
− Proceeds from local sales	0.6	0.7	1.0	8.6	8.2	0.3
Balance-of-payments impact	*14.3*	*25.1*	*35.6*	*43.4*	*45.6*	*28.9*
B. International financial flows:						
+ Proceeds from export sales	20.1	32.0	68.1	92.0	72.4	36.6
− Imports of materials	4.2	11.7	13.5	19.5	13.7	13.6
± Variation in materials in transit	(0.2)	—[a]	(2.2)	(1.0)	—	(2.3)
− Imports of services	6.4	12.3	22.2	38.5	37.5	34.8
− Net factor income paid abroad	0.2	1.0	14.2	19.4	19.0	—[a]
± Variation in overseas cash and current account balance	1.9	13.9	(1.3)	5.9	21.3	0.4
+ Reimbursement of harbor dues paid in Nigeria on behalf of the shipper	0.6	1.0	2.0	2.4	1.9	0.4
± Foreign capital flow	2.7	3.2	19.0	21.5	20.3	42.2
Balance-of-payments impact	*14.3*	*25.1*	*35.6*	*43.4*	*45.6*	*28.9*
C. Aggregation into normal balance-of-payments format:						
Visible trade:						
+ Proceeds from export sales	20.1	32.0	68.1	92.0	72.4	36.6
− Imports of materials	4.2	11.7	13.5	19.5	13.7	13.6
Trade balance	15.9	20.3	54.6	72.4	58.7	23.0
Invisible transactions:						
Trade balance	15.9	20.3	54.6	72.4	58.7	23.0
− Imports of services	6.4	12.2	22.2	38.5	37.5	34.8
− Net factor income paid abroad	0.2	1.0	14.2	19.4	19.0	—[a]
Balance on goods and services	9.3	7.1	18.1	14.6	2.2	(11.8)
Transfer payments:						
Balance on current account	9.3	7.1	18.1	14.6	2.2	(11.8)
± Capital transactions	5.0	18.0	17.5	28.8	43.5	40.7
Balance-of-payments impact	*14.3*	*25.1*	*35.6*	*43.4*	*45.6*	*28.9*

Data from unpublished individual company data. Parentheses indicate a negative figure in cases in which the entry could be positive or negative.

[a] Less than £50,000 Nigerian.

Factor Contributions to Other Sectors   77

income paid abroad, increases in overseas cash balances, and proceeds from local sales from both sides of the receipts-equal-expenditures equation. Table 6.5 contains the results of employing both methods using recent Nigerian oil industry data. In addition, part C. of this table shows the aggregation of the international financial flows into a normal balance-of-payments format.

It is of interest to consider briefly the components of the two largest items on the debit side of the petroleum industry's balance of payments, imports of services and net factor income paid abroad. To date imports of services has been the category commanding the greatest expenditure of foreign exchange. In a fairly typical year, 1965, the £22.2 million that the petroleum industry used to purchase imports of services was divided in the following manner (in million £ Nigerian): payments abroad to contractors in foreign exchange, 17.5; staff expenses, 2.2; technical advice, 1.3; insurance, 0.1; and miscellaneous, 1.1. Nearly 80 per cent of imports of services in 1965 thus consisted of payments abroad to contractors. Items included in the category of net factor income paid abroad are principally dividends, interest, and profits remitted abroad. The figures for this entry in Table 6.5 are mainly dividends remitted. In the future, of course, this category will become very large as the industry repatriates its earnings and amortizes its investment.

The balance-of-payments impacts of the petroleum industry are more interesting in comparison with the entire Nigerian balance of payments for the years in question. In Appendix Table A.2 the results of Table 6.5 are given their proper framework. The sheer bulk of information contained in Table A.2 necessitates some aggregation to provide for quick year-to-year comparisons of the petroleum sector's foreign exchange contributions. What is desired is a measure of aggregate foreign exchange availability and use with which the petroleum balance-of-payments impact can be contrasted. Accordingly, foreign exchange availability is defined in Table 6.6 as the petroleum sector's contribution (as calculated in Table 6.5) plus the sum of nonpetroleum exports, capital flows,

TABLE 6.6. FOREIGN EXCHANGE AVAILABILITY AND USE, 1963–68
(*Million £ Nigerian, except as otherwise indicated*)

Item	1963	1964	1965	1966	1967	1968
FOREIGN EXCHANGE AVAILABILITY						
From petroleum (net)[a]	14.3	25.1	35.6	43.4	45.7	28.9
From nonpetroleum[b]	170.4	220.3	255.3	208.6	175.3	211.0
Total	184.7	245.4	290.9	252.0	221.0	239.9
Petroleum as per cent of total	7.7%	10.2%	12.3%	17.2%	20.6%	12.0%
NONPETROLEUM FOREIGN EXCHANGE USE						
Imports of goods and services	227.0	261.9	281.7	262.5	258.4	247.8
Net transfers paid abroad[c]	2.7	0.2	−2.7	−1.7	−4.2	−5.8
Net increase in foreign exchange reserves[d]	−45.0	−16.7	11.9	−8.8	−33.2	−2.1
Total	184.7	245.4	290.9	252.0	221.0	239.9

Based on Appendix Table A.2.
[a] Exports plus net capital inflows minus imports minus invisible outflows.
[b] Exports plus net capital inflows plus errors and omissions.
[c] Minus indicates net inflow on transfer account.
[d] Minus indicates decrease in foreign exchange reserves.

and errors and omissions. This foreign exchange availability must equal foreign exchange use, with the latter spread among nonpetroleum imports of goods and services, net transfers paid abroad, and net increases in foreign exchange reserves. A useful indicator of the importance of the petroleum sector's contribution of foreign exchange is the petroleum balance-of-payments impact expressed as a percentage of total foreign exchange availability (or use). Table 6.6 indicates that this percentage nearly tripled between 1963 and 1967, demonstrating that the Nigerian economy has become increasingly dependent on the oil sector to provide foreign exchange.

Table 6.6 also reveals that Nigerian foreign exchange reserves have been falling steadily (with the exception of the year 1965). As mentioned earlier, this fact and the other Chenery-Strout indicators provide some evidence that Nigerian growth has been constrained by foreign exchange limitations. If these indicators are correct and Nigeria is indeed facing a foreign exchange constraint, then the petroleum industry's contributions of this scarce factor are critical for Nigerian rehabilitation and development.

In brief, the oil industry's earnings of foreign currency have made important contributions at the margin but have not yet broken Nigeria's foreign exchange bottleneck.

Indirect imports associated with the petroleum industry's local expenditures. Discussion of the foreign exchange contribution of Nigeria's petroleum sector has so far ignored indirect imports associated with the petroleum industry's local currency expenditures.[2] Only the first round of imports, i.e. the exploring and producing industry's direct purchases of imported goods and services, has been allowed for. Two methods of finding the direct and indirect imports associated with the expenditures of the Nigerian petroleum industry are demonstrated in this section. The first requires use of an input-output table and is wholly adequate for the task. The second is included to suggest a useful approach in the absence of input-output data.

An input-output table of the Nigerian economy exists for the year 1965.* One can thus calculate the A matrix whose elements, a_{ij}, refer to the amounts of input from sector i required for unit operation of sector j. As the first step in finding total imports required by the petroleum sector's operations, it is necessary to obtain the Leontief inverse matrix $(I - A)^{-1}$, from the original A matrix. In addition, information is needed on intermediate import requirements per sector of the economy in order to estimate the M diagonal matrix of import coefficients, m_j, each of which measures the amount of intermediate imports required to achieve one unit of sectoral gross output. Finally, it is necessary to know the Y vector of deliveries to final demand or net output for each of the sectors. The vector of total intermediate imports, L, can then be found by calculating the product of the matrix of import coefficients, M, the Leontief inverse matrix, $(I - A)^{-1}$, and the vector of net output, Y. The total amount of intermediate imports used by any one sector, such as the petroleum sector, is available on inspection of the L vector. If interest is focused solely on indirect

* This table is an updated and aggregated version of one originally constructed by Nicholas G. Carter for 1959–60. See Appendix D, Table D.1, for details of the updated 1965 table.

intermediate imports required by the petroleum sector, one only needs to know the import coefficients of all supplying sectors plus the amounts supplied.*

This method can be readily applied to find total imports from the Nigerian petroleum sector in 1965. By substituting new annual m_j coefficients and values for the petroleum sector's intermediate purchases, one can also calculate total imports for 1966 through 1968. Table 6.7 shows the results of this analysis. Indirect imports estimated by this method vary between £1 million and £4 million during the period under consideration.

The focus of the analysis now shifts to the second method for estimating total imports. In the absence of an input-output table it is convenient to treat second-round imports first and then to estimate third- to nth-round imports and thereby the entire import content of a given level of local expenditure of the petroleum industry. To find total imports with this method, it is necessary to trace the import content associated with oil industry purchases of goods and services backward through time. This involves treatment of rounds of expenditure beginning with that of the exploring and producing industry itself. Toward this end Table 6.8 extends discussion of the production function for petroleum from Chapter Five. Again 1965 data are employed (primarily because this is the year for which detailed survey information on servicing and supplying firms is available). As can be seen in the table, the exploring and producing industry expended a total of £69.1 million in 1965, including £35.7 million of direct imports of goods and services and £18.5 million of local goods and services. The remaining £14.9 million was made up of payments to the Nigerian

---

* This discussion can be summarized algebraically in the following manner:

(1) $L = M \cdot X$ by definition;
(2) $X = (I-A)^{-1} \cdot Y$ from Chapter Four;
(3) $L = M \cdot (I-A)^{-1} \cdot Y$ by substitution of (2) into (1);
where $L \equiv n \times 1$ vector of intermediate imports required by sector, $L_i, i = 1, \ldots, n$;
            $M \equiv n \times n$ diagonal matrix of coefficients, m_j, of intermediate import requirements by sector;
            $X \equiv n \times 1$ vector of gross output by sector;
            $I \equiv n \times n$ identity matrix;
            $A \equiv n \times n$ matrix of input-output coefficients, a_{ij}; and
            $Y \equiv n \times 1$ vector of net output by sector, $Y_i, i = 1, \ldots, n$.

TABLE 6.7. TOTAL IMPORTS ASSOCIATED WITH OPERATIONS OF THE
NIGERIAN PETROLEUM INDUSTRY, 1965–68
(*Million £ Nigerian, except as otherwise indicated*)

Imports	1965	1966	1967	1968
1. Coefficients of requirements of intermediate imports of nonpetroleum sectors (m_j, $j = 2 \ldots 4$):				
Agriculture	0.010	0.008	0.008	0.007
Manufacturing	0.214	0.182	0.161	0.147
Services	0.062	0.049	0.044	0.041
2. Total purchases by the petroleum sector from nonpetroleum sectors (X_{1j}, $j = 2 \ldots 4$):				
Agriculture	0.0	0.0	0.0	0.0
Manufacturing	10.1	18.7	17.8	7.0
Services	5.9	11.0	10.4	4.1
3. Total indirect imports associated with petroleum production $\left(\sum_{j=2}^{4} m_j X_{1j} \right)$	2.6	3.9	3.4	1.2
4. Total direct imports of the petroleum industry	35.7	58.0	51.2	48.4
5. Total direct and indirect imports of the petroleum industry	38.3	61.9	54.6	49.6

Data from Tables 6.5, D.1, D.2, D.3, D.4, and D.29; and author's estimates.

government or the Nigerian Ports Authority (NPA), local wages and salaries, and residual items, all of which have an import content only when expended. Further consideration of these expenditures is not pertinent to this discussion, since future expenditure of factor incomes is not at issue here.

The first-round imports in 1965 equaled 52 per cent of the petroleum industry's total local spending. Second-round imports are found by summing the import content of all supplying and servicing categories (the first nine lines of the second column of Table 6.8). These amounted to £5.3 million or 29 per cent of the total industry purchases of local goods and services of £18.5 million. Taking into consideration first- and second-round effects, import content was £41.0 million, or nearly 60 per cent of gross output in 1965.

The unusually low import content percentage, 29 per cent, is

TABLE 6.8. IMPORT COEFFICIENTS FOR INDUSTRIES
SUPPLYING OR SERVICING THE PETROLEUM INDUSTRY, 1965 DATA
(*Million £ Nigerian, except as otherwise indicated*)

Industry expenditure categories	Total expenditures	Import content of expenditures	Import coefficients
Local purchases of goods (including fuel and lubricants)	3.903	1.917	.491
Well-site preparation and roads	3.086	1.002	.325
Drilling	3.000	0.302[a]	.101
Transportation	2.474	0.925	.374
Construction	1.395	0.289[b]	.207
Seismic surveys	1.183	−0.135[c]	−.114
Dredging	.522	0.216	.414
Catering	.230	0.050	.217
Other local expenditures	2.711	0.783[d]	.289[d]
Wages and salaries	2.706	—	—
Payments to the Nigerian government	13.345	—	—
Harbor dues and port charges	1.997	—	—
Residual (interest, depreciation, profits, etc.)	−3.233	—	—
Imports of goods and services	35.733	35.733	1.000
Total proceeds equal total expenditures	69.052	41.082	.595

Data from Table C.1, and unpublished data for ancillary firms.

[a] Net figure from foreign exchange outflow of 0.664 and inflow of 0.362.
[b] Net figure from foreign exchange outflow of 0.439 and inflow of 0.150.
[c] Net figure from foreign exchange outflow of 0.093 and inflow of 0.228; the minus sign indicates a net inflow.
[d] Estimated on the basis of an import coefficient assumed to be equal to that of the average for all local purchases of goods and services.

in large degree the result of an institutional arrangement between the exploring and producing industry and its suppliers and contractors whereby the former pays the latter partly in freely convertible foreign currency and partly in Nigerian currency. The offshore component shows up in industry imports of goods and services and hence need not be considered in the second-round imports. But for sake of comparison, if this component is included in treatment of the ancillary firms, the import content of the industry's total (local currency plus offshore) payments to its suppliers and contractors was about 60 per cent for 1965, or £19.5 million out of £32.6 million.

The analysis so far has led to evaluations of the petroleum in-

dustry's direct import requirement, i.e. the first round, and of the major portion of the indirect requirement, i.e. the second round. The following method of approximation can be used to gain a rough order of magnitude of imports from the third and all subsequent rounds of the input-output chain of expenditures associated with the oil industry's operations. First, the value of goods and services supplied locally to the petroleum industry, £18.5 million in 1965, is multiplied by the fraction of local goods and services that is made up of domestic intermediate inputs, .124 for 1965.[3] Finally this new product is divided by a figure that equals one minus the fraction for domestic intermediate inputs.[4]

On this line of reasoning, imports from third and all subsequent rounds amounted to £0.8 million in 1965. Addition of this figure to the first- and second-round amounts already calculated places the total import requirements of the petroleum industry in 1965 at £41.8 million, or slightly over 60 per cent of the industry's expenditures. In light of second- and subsequent-round effects (assuming that these all occur during a given accounting year), the annual Nigerian petroleum industry balance-of-payments impacts might be overstated by about one-third of purchases of goods and services, or £6.1 million out of £18.5 million in 1965. The reported foreign exchange contribution of £35.6 million in 1965 nets out to £29.5 million when all post-first-round imports are considered. Including only imports from the second round, the net figure is £30.3 million. Since payments to government and to the Nigerian Ports Authority should grow much more rapidly than payments to suppliers and contractors, the adverse effects of indirect imports should become progressively less important.

Skilled Labor

In addition to supplying the two factors of production already enumerated, the Nigerian petroleum industry also makes an important though limited contribution to the development of Nigerian human resources. Initially the search for and discovery of crude petroleum attracted managers, technicians, and skilled labor to Nigeria from abroad in the form of international oil company

TABLE 6.9. LEVELS OF EMPLOYMENT IN THE NIGERIAN PETROLEUM
EXPLORING AND PRODUCING INDUSTRY, MARCH 31, 1964–67
(Number of employees)

Employees	1964	1965	1966	1967
Total employees	3,075	3,135	3,438	3,901
Nigerian employees:				
Total	2,627	2,668	2,900	3,252
Management	74	6	16	16
Professional		85	534	141
Supervisory	558	39		465
Skilled labor		1,750	882	1,043
Unskilled labor	1,393		898	969
Other	602[a]	788	570	618
Expatriate employees:				
Total	448	467	538	649
Management	301	31	39	47
Professional		280	495	379
Supervisory	147[a]	156[a]		185
Skilled technicians			4	38

Data from Nigeria [4], p. 18; [5], p. 5; [6], p. 5; [7], p. 7.
[a] Residual category.

employees. Table 6.9 pieces together the scattered bits of historical data on industry employment. In 1967 over six hundred expatriates were employed in Nigeria by the international oil companies.

In the face of local political influences as well as cost pressures, the companies have undertaken training schemes at all levels in order to employ local personnel as rapidly as feasible. The Immigration Act of 1963 placed tight controls on the number of expatriates that the exploring and producing companies and their suppliers and contractors could employ in Nigeria. The very rational aim of this policy was to force the industry to train Nigerians or to use already skilled local employees. Unfortunately the issue of expatriate quotas has been the source of a great deal of misunderstanding and frustration for both the government and the companies. Table 6.9 also gives some evidence of the steady if not overwhelmingly rapid success of the Nigerianization policy. In 1967, 622 Nigerians, nearly one-fifth of the 3,252 Nigerian employees, held high-level (management, professional, or supervisory) positions. This number was about equal to the total expatri-

ates employed at that time. Comparative data are not available on specific high-level positions held by Nigerians and expatriates, so it is not possible to contrast the relative decision-making responsibilities of the two groups.

There are basically three avenues through which the industry has trained and upgraded its Nigerian employees—scholarships, in-house training programs, and on-the-job training.[5] Industry scholarships for course work at universities and technical institutions increased from 69 in 1962 to 333 in 1966.[6] Smaller numbers of scholarships have been available for specialized secondary school training. By far the largest in-house training program was Shell-BP's Company Training School at Port Harcourt, which trained over 200 school leavers as skilled craftsmen.[7] The companies also have special internal courses for Nigerian employees, often training higher-level employees overseas. On-the-job training is naturally carried out constantly, sometimes directed by special departmental trainers, supervisory employees whose major function is training rather than line work.

Counting the expatriates attracted by oil, there is little question that the petroleum industry has been a net supplier of trained manpower to the Nigerian economy. If only skilled Nigerians are considered, the point is more debatable. The extent of petroleum sector contributions of skilled labor to the non-oil sectors of the Nigerian economy is difficult to assess. Evidence of Nigerians who have been trained by the oil industry leaving to enter other industries or to establish new enterprises is limited to a few specific instances. On the whole, the establishment of the petroleum industry in Nigeria has resulted in a fairly significant upgrading of a very limited spectrum of Nigeria's human resources. If, however, the Nigerian economy reaches a point where further growth is constrained by lack of skilled manpower, it is highly doubtful that the petroleum industry in the course of its normal operations will be able to do much about breaking this bottleneck. To date the petroleum sector's most important role in Nigerian development has been in contributing foreign exchange, Nigeria's scarce factor, and to a lesser extent in supplying investment resources.

CHAPTER SEVEN

Linkage Effects

The second category of indirect benefits associated with direct private foreign investment comprises linkage effects. As defined in Chapter Four, these are economic benefits or costs that arise as a result of intersectoral relationships. Like the direct gains estimated in Chapter Five, linkage benefits must be measured net of opportunity costs.

In the examination of linkage effects resulting from operations of the Nigerian petroleum industry, it is convenient to consider both the immediate effects of the industry's activity and fiscal linkages stemming from its payments to government. Within each of these categories there is scope for the three types of intersectoral influences described in Chapter Four. The first two of these, investment linkages (including backward and forward effects) and final demand linkages, impart benefits as a result of economies of scale, positive externalities, use of unemployed or underemployed resources, or attraction of additional private foreign investment. Costs result from negative externalities or from the creation of additional unemployment of local factors. The third, technological linkages, is a catchall category for positive or negative externalities that are not contained within either of the first two groups.

Explicit measurement of benefits and costs arising from intersectoral relationships of the Nigerian oil industry is precluded by lack of sufficiently detailed information. This dearth is partly inherent in the concepts themselves—economies of scale, externalities, opportunity costs, etc., are notoriously difficult to measure convincingly—and partly peculiar to the Nigerian scene. Much of

the evidence of linkages that is presented in this chapter is thus qualitative in nature.

Investment Linkages

Any benefits that arise from investment linkages must be considered net of opportunity costs in order to isolate the true gains associated with economies of scale, externalities, or use of unemployed or underemployed resources. Gains associated with the newly created industry's use of inputs thus have to be measured net of the social costs of factors and materials employed in creating the inputs. For example, if the Nigerian oil exploring and producing industry employs a local construction firm that would have been identically employed elsewhere in the domestic economy in the absence of petroleum, then clearly no gain to Nigeria attaches to this intersectoral relationship. In addition, gains derived from the newly created industry's supplying of its output for use as inputs by other sectors need to be taken net of the value of the output in an alternative use. For example, if natural gas in Nigeria is not exportable and is flared for lack of a domestic market, then local use of this resource can be considered nearly costless and hence a net addition to Nigerian welfare.

Backward linkages. The construction in Chapter Five of the production function for the Nigerian petroleum exploring and producing industry offered preliminary evidence that backward linkages are not likely to be strikingly significant. It will be recalled that £18.5 million or about 27 per cent of total petroleum industry expenditures in Nigerian currency was paid to suppliers and contractors operating in Nigeria. In Table 5.7 the analysis was carried one step further, and this amount was apportioned among the ancillary firms' factor recipients and material inputs. To review, the breakdown is as follows (in millions of £ Nigerian): Nigerian local materials (1.3); residual, including depreciation, interest, profits, foreign taxes, etc. (7.1); local wages and salaries (4.1); payments to government (0.9); and imported materials (5.0). These figures establish an upper limit for backward-linked benefits from petroleum that might have accrued to Nigeria

TABLE 7.1. SELECTED ANCILLARY FIRMS' RECEIPTS FROM THE PETROLEUM INDUSTRY PAID ABROAD AND PAID LOCALLY BY OWNERSHIP OF FIRMS, 1965 DATA
(Million £ Nigerian)

			Paid locally				
		Paid		Expa-triate-	Nigerian-owned		
Categories of firms	Total	abroad	Total	owned	Total	Large[a]	Small[b]
Suppliers:							
Local purchases of goods[c]	2.735	—	2.735	2.713	0.022	0.015	0.007
Contractors:							
Well-site preparation and roads	2.781	—	2.781	2.361	0.420	0.420	—
Dredging	0.470	—	0.470	0.470	—	—	—
Drilling	8.603	5.956	2.647	2.647	—	—	—
Surveys	2.044	0.956	1.088	1.088	—	—	—
Construction	3.483	2.417	1.066	0.902	0.164	0.140	0.024
Catering	0.209	—	0.209	0.150	0.059	0.059	—
Transportation	2.950	0.750	2.200	1.970	0.230	0.205	0.025
Total contractors	20.540	10.079	10.461	9.588	0.873	0.824	0.049
Total suppliers and contractors	23.275	10.079	13.196	12.301	0.895	0.839	0.056

Data from an unpublished survey of supplying and contracting firms (see Appendix C). Includes only firms interviewed in the survey.

[a] Firms with 1965 receipts from oil industry of at least £15,000.
[b] Firms with 1965 receipts from oil industry of less than £15,000.
[c] Including fuel and lubricants. Disregards instances where firms acted as manufacturer's representative for the purpose of importing goods.

in 1965. Without undertaking a detailed firm-by-firm analysis of the backward-linked enterprises, little can be said about the degree to which their relationships with the petroleum sector have led to important economies of scale or positive externalities. Since the lion's share of local expenditure by the petroleum sector involved local services rather than local goods, it is unlikely that large economies of scale were achieved.

More can be said about opportunity costs of inputs employed by the servicing and supplying firms and hence the scope of benefits stemming from use of previously underused resources. Insight here can be gained from knowledge of ownership patterns among the ancillary firms and of their activities outside the petroleum sector. Table 7.1 contains information on earnings and ownership

Linkage Effects 89

of selected firms that received payments from the Nigerian petroleum industry in 1965. The firms are aggregated into seven contracting categories and one supplier category. The petroleum-related receipts in local currency reported in the survey underlying Table 7.1 covered 88 per cent of the exploring and producing industries' onshore payments to contractors and 70 per cent of industry payments in Nigeria to suppliers (see Appendix C).

Several of the international firms, in particular those giving specialty services to the oil industry (e.g., drilling, seismic surveys), receive a substantial portion of their payments abroad.* Generally this results from a contractual arrangement whereby the expatriate servicing firm agrees to receive Nigerian currency in payment for services rendered only to the extent of its needs in Nigeria, with the remainder paid abroad in fully convertible currency. These offshore payments clearly have no impact on the Nigerian economy. But their inclusion in the discussion here does offer perspective regarding relative oil industry expenditures among overall supplying and contracting categories.

Also included in Table 7.1 are ownership data broken down among expatriate, large Nigerian, and small Nigerian categories. Because the survey upon which this table is based included an almost complete census of expatriate and large Nigerian firms but only a sampling of small Nigerian firms, the results portrayed are slightly skewed. The actual split of the £14.6 million paid by the oil industry to its contractors in 1965 was 82 per cent to expatriate-owned and 18 per cent to Nigerian-owned firms. About 88 per cent of the £3.9 million expended on supplies was purchased from expatriate firms. Overall expatriate-owned supplying or servicing firms thus undertook more than 83 per cent of oil industry business, leaving only 17 per cent for Nigerian firms.

Table 7.2 shows that most firms working for the Nigerian exploring and producing industry are not at all concerned wholly or even primarily with petroleum. Leaving aside the drilling and seismic surveying firms, all of which were employed only by the

* See the discussion of indirect imports in Chapter Six, pp. 78–83.

TABLE 7.2. SELECTED ANCILLARY FIRMS' RECEIPTS PAID LOCALLY FROM ALL INDUSTRIES AND THE SHARE FROM THE PETROLEUM INDUSTRY, 1965 DATA
(*Million £ Nigerian, except as otherwsie indicated*)

Categories of firms	Total	From petroleum industry	
		Amount	Share of total
Suppliers:			
Local purchases of goods[a]	14.874	2.735	.184
Contractors:			
Well-site preparation and roads	10.728	2.781	.259
Drilling	2.647	2.647	1.000
Transportation	7.707	2.200	.285
Construction	18.320	1.066	.058
Seismic surveys	1.088	1.088	1.000
Dredging	.970	.470	.485
Catering	.285	.209	.733
Total contractors	41.745	10.461	.251
Total suppliers and contractors	56.619	13.196	.233

Data from Table 5.4 and Appendix C.

[a] Including fuel and lubricants. Apart from interviewed fuel and lubricant suppliers whose total receipts were £12.600 million, the petroleum industry accounted for about 72 per cent of the (£2.274 million) total receipts of other interviewed local suppliers.

petroleum industry, most contracting and supplying firms did the major portion of their work within other sectors in the Nigerian economy. This was especially true of the very large expatriate construction firms and of the firms marketing petroleum products (fuel and lubricants being by far the largest local purchase of the exploring and producing industry). But without knowing the relative profitability of oil as opposed to non-oil contracts, it is impossible to establish precisely the degree to which any company's existence in Nigeria depends on the petroleum sector and hence the net gain to the Nigerian economy.

In short, the firms supplying and servicing the Nigerian oil industry are most likely to be expatriate-owned and managed, to be large relative to other firms operating in Nigeria, and with a couple of obvious exceptions to have been working in Nigeria on a large scale prior to the emergence of crude oil production. Many of the most highly specialized firms are well-established international corporations that service and supply the petroleum industry in its

Linkage Effects 91

operations all over the world. But the magnitude of work done by this group is smaller than that of the more broadly focused firms whose Nigerian experience is highly valued by the petroleum industry.

To the extent that resources employed by the ancillary firms would have found alternative local employment with equal remuneration, the gains to Nigeria are naturally reduced. The fact of large expatriate ownership seems to imply that additional foreign capital has entered Nigeria as a result of petroleum. But careful examination of the ancillary firms has shown that much of this foreign capital came into Nigeria prior to the emergence of significant activity in the petroleum sector. If these foreign-owned firms could find alternative opportunities elsewhere in Nigeria, then backward linkages from petroleum are reduced. But it is very likely that only a small minority could actually operate on the same scale apart from oil. Hence to the extent that Nigerian factors received additional incomes as a result of their employment by ancillary firms, the oil sector imparted backward linkages. Returning to the £18.5 million expended by the exploring and producing companies in 1965, it is likely that the net gain to Nigeria was something less than half of this amount. In view of the above discussion, social opportunity costs might have included amounts spent on all imported materials, nearly all local materials, perhaps half of wages, and possibly a fourth of the residual.

Several of the exploring and producing companies encourage Nigerian firms through the use of such devices as guaranteed long-term service contracts. This is especially true for firms active in well-site preparation, transportation, and construction. Within the existing supplying and servicing spectrum there is certainly scope for increased Nigerian participation, but for a variety of reasons this does not seem to be occurring very quickly. Some Nigerian would-be entrepreneurs charge that they are prevented from servicing or supplying the oil industry as much by expatriate collusion as by any lack of technical expertise, experience, or sources of finance. Over the medium term, of course, the Nigerian economy should begin to produce some of the goods and supply more

of the services that are presently imported; the much discussed but still non-existent Nigerian steel complex might be a supplier of pipe, for example. In the main, however, the nature of the petroleum industry and the state of Nigerian technology imply a very low input of Nigerian-produced materials and only a slightly greater use of Nigerian services. Currently, and for the near future, backward linkages, especially as defined net of opportunity costs, are likely to be of strictly marginal importance to the Nigerian economy.

Forward linkages. The discussion to this point has emphasized the production of crude oil in Nigeria and played down the relatively much less important fact that large quantities of associated natural gas are also produced as a joint product with crude petroleum. Both crude oil and natural gas can be used as raw materials in processing industries. In addition, natural gas is an excellent direct source of energy. Benefits associated with Nigerian firms' use of crude oil and natural gas as inputs are the subject of this section.

The Nigerian Petroleum Refining Company (NPRC) is a joint venture among the Nigerian governments (50 per cent), British Petroleum (25 per cent), and Royal Dutch Shell (25 per cent), with BP acting as the managing partner.* NPRC constructed and operated Nigeria's first and only refinery at Alesa-Eleme, near Port Harcourt in Rivers State. The refinery started operating in October 1965 and ran successfully until the outbreak of the civil war in July 1967, at which time it was forced to close down. The cost of the refinery was approximately £10.5 million, and share capital is £4 million. The oil companies involved have provided a loan of £6 million to be repaid over the first ten years of the refinery's operation. Working capital normally amounts to about £1 million. Following construction to repair the damage caused by the civil war, the refinery resumed operation in May 1970.

* The Shell-BP Petroleum Development Company of Nigeria Limited, an exploring and producing company, and the NPRC, a refining company, are wholly separate corporate entities with no direct relationship except ultimate (partial) common participation in the same vertically integrated international oil corporations.

TABLE 7.3. PETROLEUM REFINERY OPERATIONS, JANUARY–APRIL 1967
(*Long tons*)[a]

A. CRUDE AND SPECIALIZED PRODUCTS INTAKE

Item	Received	Processed
Local crude delivery	502,170	509,535
Additives imported	148	148
Total	502,318	509,683

B. FINISHED PRODUCTS

Item	Produced	Local delivery	Exported
Liquified petroleum gases	1,903	1,834	—
Premium motor spirit	53,356	56,914	—
Regular motor spirit	52,597	53,290	745
Dual purpose kerosene	81,823	77,695	158
Automotive gas oil	128,397	131,298	2,373
Fuel oil	186,277	109,648	75,299
Total	504,353	430,679	78,575

Data from Nigeria [8], p. 17.

[a] One long ton per year equals approximately .02 barrels per day; to convert data in this table from long tons per four months to barrels per day, multiply by .06.

Table 7.3 contains the most recent available information relating to refinery operations.* The rate of 31,000 barrels per day at which the refinery was processing crude in early 1967 was just sufficient to meet domestic Nigerian demand for the most important petroleum products of the refinery. Petroleum products consumed in Nigeria but not locally produced include bitumen, asphalt, lubricating oils, and aviation gasoline. Capacity of the refinery has been expanded to 55,000 barrels per day. If local demand continues to grow at about 8 per cent per annum, the refinery should be able to supply the entire local market until 1974 or 1975 with this expanded capacity.

Several international petroleum corporations have approached the Nigerian government about the possibility of constructing a new refinery of approximately the same size as the existing refinery, to be located in Lagos or in the Mid-Western State. Included

* Because these data refer to the first four months of 1967 only, reasonable estimates for normal 1967 activity can be achieved by multiplying the table figures by three.

in this group are BP and Shell, Mobil, Texaco, Compagnie Française des Pétroles (Total), Agip, and the Japan Consulting Institute.[1] It is not clear whether these companies plan to export refined petroleum products or to compete for local demand with the refinery of the Nigerian Petroleum Refining Company. The crude petroleum processed in the NPRC refinery is all produced by Shell-BP. But neither Shell-BP nor the NPRC owns the crude as it is processed. The seven major marketing companies operating in Nigeria purchase the crude from Shell-BP and then pay a refining charge to the NPRC to have it manufactured into salable petroleum products.* Annual refining charges for crude processed at a rate of 31,000 barrels per day amount to slightly less than £4 million.

Net foreign exchange benefits of the refinery are substantial. The value of the imports of petroleum products replaced by refinery production is about £14 million (at a rate of 31,000 barrels per day). The import content cost of refinery production, including imported chemicals, catalysts, and other additives, repatriation of the loan, remittance of profits, and international shipping charges amounts to about £1.8 million. Exports (mainly fuel oil for which the Nigerian demand is inadequate) carry an annual value of about £0.8 million. Under the reasonable assumption that the crude oil processed in the refinery would not otherwise be exported (demand, as set by head offices, not Nigerian supply factors, normally constrains export levels for Nigerian crude), the net foreign exchange benefits of the refinery total £13.0 million per annum.

The net contribution of the refinery to Nigerian value added can also be estimated. To produce petroleum products worth £14.0 million, the refinery might have had the following costs (in millions of £ Nigerian): crude petroleum, 8.7; imported goods and services, 1.8; and local goods and services, 1.5. All imported and most local goods and services carry a social opportunity cost equal

---

* The major marketing companies and their approximate shares of the Nigerian market are as follows: British Petroleum, 32 per cent; Shell, 24 per cent; Mobil, 18 per cent; Compagnie Française des Pétroles (Total), 8 per cent; Esso, 8 per cent; Texaco, 7 per cent; and Agip, 3 per cent.

to their full value. In addition, a portion of the £2.0 million residual is remitted abroad. If one assumes that the crude used by the refinery would not otherwise be exported, then the net gain to Nigeria from refinery operations might approach £11 million. Note, however, that since Shell-BP is paid for the crude that passes through the refinery, ultimate benefits to Nigeria can only be estimated by considering the aggregate operations of this exploring and producing company. From the results of Chapter Five, one can surmise that direct benefits will amount to perhaps one-fourth of nominal earnings. The net gain from the refinery would thus be something under £5 million. Alternatively, if the assumption is made that all crude used by the refinery is exportable, then the gain is reduced to about £2 million.

The Nigerian consumers of petroleum products, however, have been largely unaffected by the establishment of a refinery in Nigeria. Because of the small size of the Nigerian refinery, costs of refining are relatively high. The marketing companies claim that it is nearly as efficient to import petroleum products as to manufacture them in Nigeria. In any event, prices of motor spirits (gasoline) were initially reduced two pence per gallon following government pressure, but have since been raised. If the refinery had been able to operate after mid-1967, Nigerian consumers might have avoided the increase in petroleum product prices resulting from higher international freight scales in face of the closure of the Suez Canal. Why was it, then, that a much larger, more efficient refinery was not built in Nigeria, not only to satisfy local demand but also to export petroleum products? The usual and not entirely satisfactory answer from the petroleum industry is that Nigeria is not well placed on the international oil shipping routes.

The Nigerian petroleum industry made its first deliveries of natural gas for use as a source of energy in January 1963. Initial customers were a facility of the Electricity Corporation of Nigeria and the Trans-Amadi Industrial Estate, both located in Port Harcourt. From that date until the civil war began, the industry rapidly increased its sales of natural gas for energy purposes to Nigerian utilities and industries.[2] The amounts involved are still very small, however, especially in comparison with total production. Natural

TABLE 7.4. INDUSTRIES AND UTILITIES USING NATURAL
GAS AS A SOURCE OF ENERGY, 1967
(*Thousand cubic feet*)

Industry or utility	Natural gas consumed	
	Amount	Producing field
A. Utilities—Electricity Corporation of Nigeria:		
Port Harcourt	654,637	Afam
Trans Amadi	131,772	Apara
Afam	1,412,795	Korokoro
Ughelli	2,861,030	Ughelli
Total utilities	5,060,234	
B. Industries:		
Eastern Nigerian Development Corporation (Enugu)	152,691	Apara
Nigerian Petroleum Refining Company	352,849	Bomu
Textile Mills (Aba)	138,663	Imo River
Lever Brothers (Aba)	30,967	Imo River
Associated Industries (Aba)	73,112	Imo River
Nigerian Breweries (Aba)	14,626	Imo River
Glass Factory (Ughelli)	250,059	Ughelli
Total industries	1,012,967	
C. Grand total	6,073,201	

Data from Nigeria [8], p. 16.

gas occurs either in isolation or in association with deposits of crude petroleum. In order to produce crude oil, associated gas must be either transported away or flared, since in general it is uneconomic to store. All natural gas production to date in Nigeria has been associated with crude production. Since natural gas would not be exportable from Nigeria without a considerable additional investment in processing facilities, this natural resource can be considered to have an opportunity cost approaching zero.

Table 7.4 displays information regarding the Nigerian utilities and industries that were consumers of natural gas for energy purposes in 1967. Approximately 83 per cent of the 1967 local sales of natural gas went to the Electricity Corporation of Nigeria. The remaining 17 per cent was spread among the refinery, a textile mill, a soap factory, a brewery, a glass factory, and other assorted enterprises. All consumers located in the Eastern States—the entire group less Electricity Corporation of Nigeria (Ughelli) and Glass Factory (Ughelli)—stopped receiving natural gas shipments

in July 1967. During a normal year, it might be expected that total consumption of natural gas might have amounted to 10 billion cubic feet. The value of this volume of sales would have approached £1.4 million. The producing companies themselves consumed natural gas as fuel in amounts that worked out to 5 per cent of local sales, which in a normal year might have added another £70,000 to the value of natural gas utilization. Considering the opportunity costs of alternative fuels as well as the net contribution to local factors of production associated with this aspect of the petroleum industry's operations, the annual net gain to Nigeria from this forward linkage might be about £1.0 million at 1967 consumption levels.

To place the consumption figures of Table 7.4 in perspective, it is instructive to compare them with natural gas production. Production in 1967 was 91.6 billion cubic feet. Consumption of natural gas was therefore less than 7 per cent of total production. The residual was flared and hence served no economic purpose. Moreover, as the production of crude oil expands in the future, so will output of associated natural gas. And the large proved reserves of nonassociated natural gas are still wholly untapped. There is great potential for further use of natural gas for energy purposes as well as for raw material inputs into gas-based industries.

Final Demand Linkages

It is very difficult to do more than indicate the scope that the Nigerian petroleum industry might have for stimulating benefits from final demand linkages until extensive data are available. Payments made by the Nigerian exploring and producing industry that might lead to final demand linkages include wages and salaries and residual payments (profits, interest, depreciation, etc.).*
In addition, the industry makes payments in local currency to its suppliers and contractors. Since these ancillary firms also have wage bills and residual payments, they too contribute to possible final demand linkages, though there may be a lag in the timing of their effects. Although it is important conceptually to differen-

* See Tables 5.5 through 5.7 for relevant data on industry payments during 1963–68 and for definition of the residual.

tiate between these two sources, it is convenient to consider them together in the empirical discussion that follows.

In 1965, total residual earnings were £4 million, including a negative figure of £3.3 million for the exploring and producing industry and a positive £7.3 million for the ancillary firms. War effects have caused the residual category to become strongly negative, reflecting the cutback in production of crude, which was much more severe than the reduction in the industry's local expenditures. Recently there has thus been no scope for possible final demand linkages from this source. But in the near future this situation should be reversed as the accounts of the exploring and producing companies move rapidly into the black.

The entire exploring and producing industry and about five-sixths of the ancillary firms are foreign-owned. Hence most residual earnings are likely to be transferred abroad, without benefit to the Nigerian economy. Political pressures might induce the exploring and producing companies to make certain unusual expenditures and investments in Nigeria, but for the most part profits and other residual payments will probably continue to be expatriated (after this category turns positive).

Wages and salaries hold greater promise. The year 1965 is the only one for which data exist for the supplying and servicing firms; hence it is assumed to be representative for purposes of this discussion. In 1965 the exploring and producing industry employed 3,158 Nigerians and paid them a total of £1,363,000 as wages and salaries, an average of £432 per man. Supplying and servicing firms hired 14,051 Nigerians and paid them wages of £2,284,000, or £163 per capita.* The difference in the average wage can be attributed to the fact that over one-sixth of the Nigerians employed by the exploring and producing industry were high-level employees (see pp. 83–85).

During this same year, the exploring and producing industry employed 545 expatriates and paid them (in Nigerian currency) a total of £1,343,000. This average of £2,464 per head is low be-

* For a breakdown of wages among contractor categories, see Table C.1 in Appendix C.

cause many of the companies pay significant portions of their expatriate salaries into foreign banking accounts. Concurrently, the petroleum industry's suppliers and contractors employed 1,029 expatriates. No precise data exist for these expatriates' salaries (see Appendix C). If one can assume that the expatriates employed by suppliers and contractors received the same average payment in Nigerian currency as their counterparts in the exploring and producing industry, then local salaries for the former might have amounted to about £2,535,000. Summing up, the total amount of locally paid wages and salaries generated by the production of crude oil in Nigeria in 1965 was about £7,525,000. This amount was paid to 17,309 Nigerians and 1,574 expatriates.

It is interesting to compare these amounts with the wider Nigerian picture. Employment data for Nigeria are sketchy at best. The Federal Ministry of Labour does carry out an annual employment and earnings inquiry, but its coverage is incomplete for small-scale firms. The reported total of gainfully employed wage or salary earners in Nigeria at the end of 1964 was 562,000; this may be an undercount of some 100,000 or more.[3] Of this total some two-fifths were in government, with the majority of the rest spread fairly evenly among manufacturing, construction, transport, and commerce. Comparable earnings data are available for the year 1962; the average wage was £186, with a high of £275 in commerce and a low of £102 in agriculture.[4] Data for 1965 are at hand for the Nigerian manufacturing sector; 91,768 employees received a total of £20,456,000 in wages and salaries, an average payment of £223 per man.[5] The average payment of the exploring and producing industry to its Nigerian employees is therefore significantly higher than that of other sectors, whereas the servicing and supplying firms pay somewhat below average. The latter result is not surprising, considering the large numbers of common laborers employed in drilling, seismic, and construction crews. Taken as a whole, the consolidated Nigerian oil industry pays an average wage comparable to that of the manufacturing sector—£212 and £223, respectively.

The question remains whether £7,525,000 in wages and salaries

is likely to have an important impact on the Nigerian economy. Skilled laborers received about two-thirds of the total. If their opportunity costs are taken as 100 per cent, the scope for possible final demand linkages is reduced to perhaps £2.5 million. Unfortunately it is not possible to estimate what portion of this figure leaks out as import content, since applicable budget studies showing import and savings propensities do not exist. A few brief comments are relevant, however. Because large portions of expatriate salaries are paid overseas (and are not considered in the above figures), the likelihood of sizable expatriate remittances is greatly lessened. The foreigners, however, tend to consume large amounts of imported goods with their Nigerian currency payments. It is equally difficult to speculate about the propensities of Nigerians to import and save. Compared with a total GDP figure in 1965 of £1.3 billion, or even of a nonsubsistence GDP figure of £0.9 billion, the oil industry's wage and salary payments seem infinitesimal. Although there is little evidence, wage earners in the oil industry appear to demand the types of consumer goods that import-substituting Nigerian industry is beginning to make in increasing amounts. If so, the Nigerian petroleum industry does contribute very small final demand linkages at the margin.

Technological Linkages

As defined in Chapter Four, the effects of technological linkages are externalities, i.e., nonappropriated benefits or costs arising when the operations of the industry in question supply free inputs into other industries' production functions. Examples cited there included training labor, promoting mass education, introducing new technologies, and building infrastructure. In all instances the activity of the given industry would supply intended or unintended gratuitous benefits or costs to other industries.

With respect to the Nigerian petroleum industry, one important technological linkage, the training of labor and management for subsequent employment in another industry, has already been discussed in Chapter Six (pp. 83–85). This factor contribution also has the attributes of a linkage effect. In light of southern Nigeria's

rapid move into mass education, there was no room or need for the oil industry to promote this effort even if it were somehow so disposed. As for new technologies or methods of organization, there is some indication that several Nigerian firms, especially those backward-linked with the oil industry, are making small changes in such things as bookkeeping and management practices. But no great shock effects have emerged in this regard.

Turning to infrastructure, the picture is again mixed. Most significantly, the petroleum industry has constructed and maintained several miles of roads in the Nigerian bush in the process of opening up well-sites. Qualitative evidence exists that some of these at least have been used as farm-to-market feeder roads by Nigerian farmers. Too much should probably not be made of this. But the fact that oil companies do build roads is often overlooked by certain observers who emphasize that crude petroleum is transported by pipelines and conclude wrongly that the oil industry thus imparts no external economies in transportation. Some improvement of Nigerian ports has accompanied expansion of the oil industry, but this has not significantly benefited the non-oil economy. The oil companies have built power stations but not for use outside of the oil sector. These structures, together with the social overhead capital (buildings, hospitals, schools, etc.) constructed by the oil companies for the use of their employees, if not totally depreciated, will eventually transmit externalities whenever the oil companies leave and turn them over to other sectors of the Nigerian economy. In sum, the petroleum industry in Nigeria has not communicated large technological linkages, with the two rather important exceptions of training labor and management and constructing roads.

Fiscal Linkages

Fiscal linkages are external effects associated with the government's use of revenues received from the petroleum industry. An incremental pound of revenue accruing to the government as a result of increased activity in the petroleum sector might bring about two different types of benefits. Assuming a constant level

of government expenditure, the extra revenue would allow the government to reduce tax burdens on other parts of the economy. Alternatively, assuming an unchanged level of nonpetroleum taxation, the additional revenue would permit the government to increase expenditures. Benefits that arise from either action must be considered net of collection costs. Treatment of these effects is necessarily complementary to the section of Chapter Six in which payments to government are dealt with in a factor constraint framework as a contribution to public domestic saving and thus investment resources (see pp. 71–74).

A complete evaluation of fiscal linkages would entail an examination of all Nigerian public finance, since it is not realistic to consider petroleum-related revenues and government expenditure of them separately from other revenues and expenditures. Such an evaluation would involve a detailed analysis that is not attempted in this study. Revenues and expenditures of the exploring and producing industry have been presented in Chapter Six. Total payments to the Nigerian government associated with the production of crude petroleum are summarized here. Speculation about wider effects and policy implications of fiscal linkages is reserved for the final chapter.

In the earlier handling of payments to the Nigerian government, only the payments made directly by the exploring and producing companies were included. Table 7.5 adds company taxes from the supplying and servicing firms as well as pay-as-you-earn (PAYE) withholding income taxes for the Nigerian and expatriate em-

TABLE 7.5. TOTAL PAYMENTS TO GOVERNMENT ASSOCIATED WITH THE PRODUCTION OF NIGERIAN PETROLEUM, 1965
(*Thousand £ Nigerian*)

Item	Company taxes[a]	Nigerian PAYE[b] taxes	Expatriate PAYE[b] taxes	Total
Exploring and producing industry	13,345	73	487	13,905
Supplying and servicing firms	1,258	75	475	1,808
Total	14,603	148	962	15,713

Data from Tables 6.2, C.1, and unpublished individual company data.
[a] Including all payments to government by the exploring and producing companies.
[b] PAYE, or pay-as-you-earn withholding taxes.

Linkage Effects

ployees of the consolidated industry. As indicated in Table 7.5, the total base for potential fiscal linkages in 1965 increases by nearly 13 per cent with these additions. Moreover, company taxes paid by supplying and servicing firms should grow markedly in the near future as carried-over capital allowances are used up and tax holidays expire. Quite naturally, however, the payments to the Nigerian government by the exploring and producing industry will continue to be by far the most important component of oil-related revenues.

CHAPTER EIGHT

Petroleum in Nigeria: Some Projections

In the preceding three chapters, the analytical framework of Chapter Four has been applied to the recent history of petroleum in Nigeria. This historical experience is now used as the basis for construction of a projective input-output model to analyze likely future domestic impacts of Nigerian petroleum. The empirical analysis of this chapter is based on the theoretical discussion of a requirements analysis employing input-output techniques in Chapter Four (pp. 50–54).

Magnitudes of all of the variables discussed below for use in the input-output model depend ultimately on levels of petroleum output. Two projections of future levels of Nigerian crude petroleum production and export are made—one believed to be somewhat conservative and the other quite optimistic. These projections provide the basis for speculative analyses of possible medium-term impacts of oil on the Nigerian economy. In Chapter Nine, a separate projection is made to analyze hypothetical impacts of petroleum on Biafra and thereby to gain insight into Nigerian and Biafran economic motivations in the civil war.

Production and Exports

Conditions are currently too uncertain in Nigeria to allow a causal analysis running directly from factors affecting future levels of Nigerian crude output to projections of production. A complicated mixture of factors enters into the decisions of head offices of international petroleum corporations concerning production levels in any given source of supply. Among the most important

Petroleum in Nigeria: Some Projections 105

of these are international demand for petroleum products, political and economic outlooks in all producing countries, production costs and the return on capital within the company's vertically integrated framework, financial arrangements with the government and expectations regarding their stability, and special features pertaining to crude production in the area (e.g. location and crude quality).

The buoyancy of international demand for petroleum products should continue. Attention thus centers on head office determination of relative acceleration of output in various producing areas. On the whole, the international petroleum industry is optimistic about prospects in Nigeria, the recent civil war and the changes in financial arrangements with government notwithstanding. These negative factors seem to be more than compensated by Nigeria's special advantages (location, crude quality, etc.). The industry has thus undertaken a rapid reentry into those parts of Nigeria in which productive activity was for a long time precluded by the civil war. Unlike most other private foreign investors, the oil industry seems to be nearly immune to the usual considerations of investment climate.

Table 8.1 sets out historical data and two alternative projections for production, exports, and local sales of Nigerian crude. Underlying the projections is an assumption that the current financial arrangements between the petroleum companies and the Nigerian government remain unchanged throughout the period under consideration. The high projection assumes that the industry will expand very rapidly in Nigeria, whereas the low projection is based on a somewhat more conservative growth outlook. The head offices of the petroleum exploring and producing companies make production plans for a minimum of five years ahead. The precise annual levels of output under both projections are aggregations of individual company plans. Although it is true that company head offices alter their plans periodically, the general orders of magnitude of these projections seem reasonable.[1] When most of the Nigerian oil industry was forced to shut down in July 1967, production was 600,000 barrels per day. The low projection works

TABLE 8.1. PETROLEUM PRODUCTION, 1963–73
(Volume in thousand barrels per day; value in million £ Nigerian)

A. HISTORICAL

Production	1963	1964	1965	1966	1967	1968
Volume:						
Production	76	120	270	415	317	142
Exports	76	120	266	383	300	142
Local sales	—	—	4	32	17	—
Value:						
Production	20	32	69	100	77	37
Exports	20	32	68	92	72	37
Local sales	—	—	1	8	5	—

B. PROJECTIONS

Production	1969	1970	1971	1972	1973
		LOW			
Volume:					
Production	540	1,050	1,250	1,525	1,800
Exports	540	1,023	1,214	1,486	1,758
Local sales	—	27	36	39	42
Value:					
Production	133	275	325	395	465
Exports	133	266	313	382	451
Local sales	—	9	12	13	14
		HIGH			
Volume:					
Production	540	1,200	1,575	2,000	2,400
Exports	540	1,173	1,539	1,961	2,358
Local sales	—	27	36	39	42
Value:					
Production	133	313	406	513	616
Exports	133	304	394	500	602
Local sales	—	9	12	13	14

Data from Table 5.1 and author's estimates.

out to a tripling of the July 1967 level by 1973. Alternatively the high projection foresees a quadrupling of this average output in the same time span. Production averaged less than 150,000 barrels per day during 1968, but began to recover toward the end of that year. Owing to damages and interruptions caused by war disturbances in some of Shell-BP's producing areas in the Eastern

States and to marketing difficulties experienced by Gulf, production levels fluctuated somewhat during 1969, reaching a low monthly average of about 450,000 barrels per day in August. The annual average is provisionally estimated as 540,000 barrels per day for that year. Following the end of the war in January 1970, output rose rapidly and surpassed one and a quarter million barrels per day in September 1970.

Other Local Payments

Other local payments, i.e. local currency expenditures of the petroleum industry less payments to the Nigerian government, comprise three categories of interest—local goods and services, local wages and salaries, and harbor dues and port charges. Payments in the first two instances include all payments in Nigerian currency to expatriates and expatriate-owned firms and to their Nigerian counterparts. Harbor dues and port charges go entirely to the Nigerian Ports Authority. Table 8.2 summarizes recent payments in these categories as well as the low and high projections that might be associated with the two production forecasts.

The very gently rising levels foreseen for local goods and services and for local wages and salaries (in spite of rapidly expanding production levels) reflect the unusually low costs of production or high valuation of output that are part of the curious institutional framework of the international oil industry (see pp. 7–9).[2] Within the totals for local goods and services, it is expected that payments to drilling firms will more than double the average growth rate for all local goods and services, that payments to seismic surveying firms will level off and show little if any growth, and that payments to all other contracting categories will approximate the average. Input coefficients of the petroleum production process should change considerably, with local coefficients falling as output rises.

The wages and salaries shown in Table 8.2 apply only to the exploring and producing companies. The skilled labor component should remain at about two-thirds of the total. Naturally the local

TABLE 8.2. OTHER LOCAL PAYMENTS, 1963–73
(Million £ Nigerian)

A. HISTORICAL

Payment	1963	1964	1965	1966	1967	1968
Local goods and services	7	10	19	28	22	11
Local wages and salaries	2	2	2	3	3	—[a]
Harbor dues and port charges	1	1	3	3	2	2
Total other local payments[b]	10	14	23	33	27	13

B. PROJECTIONS

Payment	1969	1970	1971	1972	1973
LOW					
Local goods and services	26	30	32	32	32
Local wages and salaries	4	4	4	4	4
Harbor dues and port charges	3	7	9	11	13
Total other local payments[b]	33	41	45	47	49
HIGH					
Local goods and services	26	35	40	40	40
Local wages and salaries	4	4	4	4	4
Harbor dues and port charges	3	9	11	14	17
Total other local payments[b]	33	47	56	59	62

Data from Table 5.2 and author's estimates. Other local payments include local currency expenditures less payments to the Nigerian government.

[a] Less than £500,000 Nigerian.
[b] May not equal sum of components because of rounding.

suppliers and contractors will also use a portion of their receipts from the exploring and producing companies to make local wage or salary payments. By 1973 the consolidated crude oil industry in Nigeria, i.e. the petroleum industry plus the ancillary firms, might employ about 23,000 Nigerians and 2,100 expatriates and pay on the order of £10 million in wages and salaries.

Harbor dues and port charges can be estimated directly from export projections. Unless the existing high rates are reduced to approximate international standards, the amounts involved should become increasingly significant. For the most part, these payments can be considered increments to the totals for payments to government, since they are made to a quasi-governmental agency, the Nigerian Ports Authority, and since the services performed by the NPA on behalf of the petroleum industry probably should be valued at less than one-third of the amounts actually

paid for them. The amounts given in Table 8.2 are understated to the extent of payments made as a result of the Offshore Terminal Dues Decree of May 1969; because no rates for offshore production have yet been announced, it is necessary to exclude this item from consideration (see Appendix B).

Payments to the Nigerian Government

The Nigerian petroleum industry contributes public domestic saving and thus investment resources to the Nigerian economy via large and increasing payments to government. Table 8.3 contains data for historical payments to government by category as well as amounts that might be associated with the low and high projections. Though the table is brief, the calculations underlying

TABLE 8.3. PETROLEUM INDUSTRY PAYMENTS TO GOVERNMENT, BY CATEGORY, 1963–73
(*Million £ Nigerian*)

A. HISTORICAL

Payment	1963	1964	1965	1966	1967	1968
Petroleum profits tax	—	5	2	3	6	4
Royalties	2	2	5	8	15	3
Other[a]	3	5	7	8	7	10
Total payments to government[b]	5	12	13	19	27	16

B. PROJECTIONS

Payment	1969	1970	1971	1972	1973
		LOW			
Petroleum profits tax	3	31	65	85	108
Royalties	6	21	33	41	49
Other[a]	14	16	18	18	19
Total payments to government[b]	23	68	116	144	176
		HIGH			
Petroleum profits tax	3	40	85	115	156
Royalties	6	24	41	53	65
Other[a]	14	16	18	18	19
Total payments to government[b]	23	80	144	186	240

Data from Table 6.2 and author's estimates.
[a] Rentals, customs and stamp duties, premiums, and other taxes.
[b] May not equal sum of components because of rounding.

TABLE 8.4. THE IMPACT OF PETROLEUM REVENUES ON TOTAL
GOVERNMENT REVENUES, 1963–73
(*Million £ Nigerian, except as otherwise indicated*)

A. HISTORICAL

Revenue	1963	1964	1965	1966	1967	1968
Petroleum revenues[a]	5	12	13	19	27	16
Nonpetroleum revenues[b]	142	166	177	178	136	140
Total revenues[b]	147	178	190	197	163	156
Petroleum revenues as per cent of total revenues	3	7	7	9	17	10

B. PROJECTIONS

Revenue	1969	1970	1971	1972	1973
		LOW			
Petroleum revenues	23	68	116	144	176
Nonpetroleum revenues	150	165	180	190	200
Total revenues	173	233	296	334	376
Petroleum revenues as per cent of total revenues	13	29	39	43	47
		HIGH			
Petroleum revenues	23	80	144	186	240
Nonpetroleum revenues	150	165	180	190	200
Total revenues	173	245	324	376	440
Petroleum revenues as per cent of total revenues	13	33	44	49	55

Data from Tables 6.2, 6.3, 8.3, and author's estimates.
[a] Calendar year.
[b] Fiscal year (starting April 1).

its content are long, tedious, and of insufficient interest to be included here.* The numbers shown are aggregations of amounts calculated on a company-by-company basis, employing estimates of each producing company's output, costs, and capital allowances.

The importance of these accelerating payments to government is clarified with reference to total government revenues. In Table 8.4 a comparison of petroleum and nonpetroleum revenues is made. No great faith is attached to the projection of nonpetroleum revenues; it is relatively conservative and assumes that prewar levels are regained by 1971. Using the low projection, 1973 total revenues are nearly 190 per cent of the 1966 peak, with oil

* See Appendix B for methodology to be employed in calculating revenues.

revenues moving from one-tenth to almost one-half of the rising total. Alternatively, on the basis of the high projection during this same period, total revenues more than double, whereas oil revenues increase to 55 per cent of the 1973 total.

Government revenues of these magnitudes will almost certainly provide increases in public domestic saving in amounts that must depend on future levels of government current expenditures. The analysis of Chapter Six (pp. 72–74) has indicated that current spending grew at 11 per cent per year between 1959 and 1966. If this high growth rate or one even higher is resumed, perhaps reflecting increased administrative expenses of the twelve-state system, rehabilitation and reconstruction expenditures, greater spending on military and ex-military personnel, etc., then the Nigerian governments could consume a substantial portion of expected total revenues under either projection. But recurrent spending on this order would surely cause the nonpetroleum revenues to grow much more rapidly than is allowed for in Table 8.4, since the major item in this category consists of customs and excise payments. And hopefully the Nigerian governments will be able to exercise prudence in their recurrent spending policies. The inability to make firm guesses about future government spending activity allows no conclusion other than that petroleum revenues should offer a tremendous potential for large and increasing contributions to public domestic saving. If one can make arbitrary assumptions about future levels of government spending, the importance of the petroleum sector's contribution of investment resources can be gauged within the context of the input-output model developed below.

Contributions to the Nigerian Balance of Payments

After supplying its own requirements, the petroleum sector furnishes substantial amounts of foreign exchange to the rest of the Nigerian economy. It will be remembered that the balance-of-payments impact of the Nigerian petroleum industry, the amount of free foreign exchange supplied to other Nigerian sectors, can be estimated through the use of two approaches, one dealing with the industry's local currency expenditures and the other with its

TABLE 8.5. BALANCE-OF-PAYMENTS IMPACT OF THE PETROLEUM
INDUSTRY, 1963–73: LOCAL CURRENCY EXPENDITURES
(*Million £ Nigerian*)

A. HISTORICAL

Payment	1963	1964	1965	1966	1967	1968
Payments to government	5	12	13	19	27	16
Other local payments	10	14	23	33	27	13
Proceeds from local sales	1	1	1	9	8	—ᵃ
Balance-of-payments impactᵇ	14	25	36	43	47	29

B. PROJECTIONS

Payment	1969	1970	1971	1972	1973
		LOW			
Payments to government	23	68	116	144	176
Other local payments	33	41	45	47	49
Proceeds from local sales	1	9	12	13	14
Balance-of-payments impactᵇ	55	100	149	178	211
		HIGH			
Payments to government	23	80	144	186	240
Other local payments	33	47	56	59	62
Proceeds from local sales	1	9	12	13	14
Balance-of-payments impactᵇ	55	118	188	232	288

Data from Tables 6.5, 8.2, 8.3, and author's estimates.
ᵃ Less than £500,000 Nigerian.
ᵇ May not equal sum of components because of rounding.

international financial flows. These two approaches are employed in Tables 8.5 and 8.6 to calculate past and possible future balance-of-payments impacts of the oil industry.

The projections in Table 8.5 draw on forecasts made in the two previous sections as well as on an entry for proceeds from local sales. Local proceeds are determined principally by crude sales to the Nigerian refinery but include also earnings from sales of natural gas and other incidental local proceeds.* Table 8.6 relies on projections of international financial flows associated with the petroleum industry, including entries for exports, imports of goods and of services, net factor income paid abroad, and net foreign

* The figures in Tables 8.5 and 8.6 overlook the net balance-of-payments impacts of the refinery. The inclusion of these would increase effective foreign exchange availability by £9 million in 1970 (since the refinery went back into operation in May 1970), £14 million in 1971, £15 million in 1972, and £16 million in 1973. See Chapter Seven, pp. 92–95.

TABLE 8.6. BALANCE-OF-PAYMENTS IMPACT OF THE PETROLEUM
INDUSTRY, 1963–73: INTERNATIONAL FINANCIAL FLOWS
(*Million £ Nigerian*)

A. HISTORICAL

Payment	1963	1964	1965	1966	1967	1968
Proceeds from export sales	20	32	68	92	72	37
Imports of materials	4	12	14	20	14	14
Imports of services	6	12	22	39	38	35
Net factor income paid abroad	—[a]	1	14	19	19	—[a]
Net foreign capital flow	5	18	18	29	44	41
Balance-of-payments impact[b]	14	25	36	43	46	29

B. PROJECTIONS

Payment	1969	1970	1971	1972	1973
		LOW			
Proceeds from export sales	133	266	313	382	451
Imports of materials	23	25	23	22	22
Imports of services	46	50	46	44	44
Net factor income paid abroad	74	131	133	173	207
Net foreign capital flow	65	40	38	35	33
Balance-of-payments impact[b]	55	100	149	178	211
		HIGH			
Proceeds from export sales	133	304	394	500	602
Imports of materials	23	27	29	27	27
Imports of services	46	54	58	53	53
Net factor income paid abroad	74	154	165	232	276
Net foreign capital flow	65	49	46	44	42
Balance-of-payments impact[b]	55	118	188	232	288

Data from Tables 6.5, 8.2, and 8.3, and author's estimates.
[a] Less than £500,000 Nigerian.
[b] May not equal sum of components because of rounding.

capital flows. As indicated in Table 8.6, foreign exchange earnings from petroleum in 1973 are over four times their 1967 level using the low projection and six times this level on the basis of the high projection. Allowing for indirect imports might reduce the contribution of the local goods and services portion of other local payments to the overall impact by as much as one-third. But this would lower the total impact in any single year by at most 15 per cent in the beginning of the projection period and 5 per cent at the end.

To confirm the important concept that the balance-of-payments impacts and not the export earnings of the petroleum industry

TABLE 8.7. BALANCE-OF-PAYMENTS IMPACT OF THE PETROLEUM INDUSTRY AS PER CENT OF PETROLEUM EXPORT EARNINGS, 1963–73
(Million £ Nigerian, except as otherwise indicated)

A. HISTORICAL

Payment	1963	1964	1965	1966	1967	1968
Balance-of-payments impact	14	25	36	43	47	29
Exports	20	32	68	92	72	37
Balance-of-payments impact as per cent of exports	70	78	53	47	65	78

B. PROJECTIONS

Payment	1969	1970	1971	1972	1973
		LOW			
Balance-of-payments impact	55	100	149	178	211
Exports	133	266	313	382	451
Balance-of-payments impact as per cent of exports	41	38	48	47	47
		HIGH			
Balance-of payments impact	55	118	188	232	288
Exports	133	304	394	500	602
Balance-of-payments impact as per cent of exports	41	39	48	46	48

Data are from Tables 8.1, 8.5, and 8.6.

have significance for the Nigerian economy, these two variables are compared in Table 8.7. After all of the quirks resulting from the Nigerian oil legislation (especially accumulated capital allowances) have worked themselves out, the annual oil balance-of-payments impact will equal nearly one-half of petroleum export earnings. Or, what amounts to the same thing, the petroleum exploring and producing companies will bring about half of their export earnings back into Nigeria.

The contributions of the petroleum industry to Nigerian foreign exchange are of major interest in the context of overall Nigerian foreign exchange availability and use. In Table 8.8 the recent Nigerian foreign exchange situation is reviewed and a plausible picture is portrayed for the medium-term future using both low and high projections. Underlying the entry entitled nonpetroleum (exports plus net capital flows), which is identical for both projections, is a forecast of all non-oil balance-of-payments en-

TABLE 8.8. PETROLEUM INDUSTRY CONTRIBUTION TO NIGERIAN FOREIGN
EXCHANGE AVAILABILITY AND USE, 1963–73
(Million £ Nigerian, except as otherwise indicated)

A. HISTORICAL

Item	1963	1964	1965	1966	1967	1968
Petroleum balance-of-payments impact	14	25	36	43	47	29
Nonpetroleum[a]	170	220	255	209	174	211
Total foreign exchange availability[b]	185	245	291	252	221	240
Nonpetroleum imports of goods and services	227	262	282	263	258	249
Net transfer paid abroad[c]	3	—	−3	−2	4	−6
Net increase in foreign exchange reserves	−45	−17	12	−9	33	−2
Total foreign exchange use	185	245	291	252	221	240
Petroleum balance-of-payments impact as per cent of foreign exchange availability or use	8	10	12	17	21	12

B. PROJECTIONS

Item	1969	1970	1971	1972	1973
LOW					
Petroleum balance-of-payments impact	55	100	149	178	211
Nonpetroleum[d]	230	250	265	275	285
Total foreign exchange availability[e]	285	350	414	453	496
Petroleum balance-of-payments impact as per cent of foreign exchange availability or use	19	29	36	39	43
HIGH					
Petroleum balance-of-payments impact	55	118	188	232	288
Nonpetroleum[d]	230	250	265	275	285
Total foreign exchange availability[e]	285	368	453	507	573
Petroleum balance-of-payments impact as per cent of foreign exchange availability or use	19	32	42	46	50

Data from Tables 6.6, 8.5, and 8.6, and author's estimates.
[a] Exports plus net capital flows plus errors and omissions.
[b] May not equal sum of components because of rounding.
[c] Dash indicates between + £500,000 and − £500,000.
[d] Exports plus net capital flows.
[e] Total foreign exchange availability equals total foreign exchange use, which includes nonpetroleum imports of goods and services, net transfer paid abroad, and net increase in foreign exchange reserves.

tries except imports of goods and services and transfer payments. The most important assumption attached to this forecast is that the value of non-oil exports, projected item by item, will grow at an aggregate rate of 3¾ per cent per annum from 1969.

It may be useful to reconsider briefly the meaning to be gained

from the format and content of Table 8.8. Foreign exchange generated by the Nigerian oil industry is measured by the petroleum balance-of-payments impact, which, it will be remembered, is net of all foreign exchange needs of the oil industry. This foreign exchange is thus available for use by the non-oil sectors of the Nigerian economy. At the same time, these other sectors generate amounts of foreign exchange of their own, as indicated in Table 8.8 by the line Nonpetroleum (exports plus net capital flows). Summing these two items gives the total amount of foreign exchange that is available to the Nigerian non-oil economy during a given year. This foreign exchange can be used to purchase nonpetroleum imports of goods and services, to pay net transfers abroad, or to build up foreign exchange reserves. The importance of the petroleum sector's contribution of foreign exchange can therefore be gauged by comparing the balance-of-payments impact of petroleum with the total foreign exchange availability or use during any given year. Under the low projection, foreign exchange availability regains its 1965 peak in 1969 and then increases by 1973 to nearly three-fourths again as much as the 1965 level. The growth of foreign exchange availability using the high projection is even more spectacular; the 1965 peak level is also attained by 1969, but the 1973 amount is about twice as large as that of 1965. The importance of the contribution of foreign exchange to future Nigerian output and income is considered below with the aid of an input-output model.

Value Added of the Petroleum Industry

Much of the projected information can be drawn on to indicate the direct contribution of the Nigerian petroleum exploring and producing industry to Nigerian gross product, using first domestic and then national income accounting concepts. This compilation begins in Table 8.9, where the industry value added, historical and projected, is calculated on both bases. Petroleum value added to Gross Domestic Product in 1973 is nearly eight times its level in 1967 of £51 million based on the low projection and over ten times this level using the high projection. The petroleum

TABLE 8.9. VALUE ADDED OF THE PETROLEUM INDUSTRY, 1963–73
(Million £ Nigerian)

A. HISTORICAL

Item	1963	1964	1965	1966	1967	1968
1. Value added to GDP:						
Calculated as a residual:						
(1) Proceeds from export sales	20	32	68	92	72	37
(2) Proceeds from local sales	1	1	1	9	8	—[a]
(3) Net foreign capital flows	5	18	18	29	44	41
(4) Imports of materials	4	12	14	20	14	14
(5) Imports of services	6	12	22	39	38	35
(6) Local goods and services	7	10	19	28	22	11
(7) Value added to GDP[b] (1)+(2)+(3)−(4)−(5)−(6)	8	17	32	44	51	18
Calculated from individual components:						
(8) Payments to government	5	12	13	19	27	16
(9) Harbor dues and port charges	1	1	2	3	2	—[a]
(10) Local wages and salaries	2	2	3	3	3	2
(11) Net factor income paid abroad	—[a]	1	14	19	19	—[a]
(12) Value added to GDP[b] (8)+(9)+(10)+(11)	8	17	32	44	51	18
2. Value added to GNP:						
Calculated as a residual:						
(13) Value added to GNP[b] (1)+(2)+(3)−(4)−(5)−(6)−(11)	8	16	18	24	32	18
Calculated from individual components:						
(14) Value added to GNP[b] (8)+(9)+(10)	8	16	18	24	32	18

(Table continued on p. 118)

sector's value added to GNP in 1973 is projected to be over six times its level in 1967 of £32 million (low projection), or more than eight times this amount (high projection).

To provide a basis of comparison, Tables 8.10 and 8.11 contain projections of nonpetroleum value added based on an arbitrary assumption of a 2.2 per cent rate of growth in 1969 and a 4.0 per cent rate thereafter. In Table 8.10 petroleum value added is compared with past and projected non-oil value added, and its contribution to Nigerian GDP is given annually as a percentage of the total. From the 1969 base, which is depressed by the civil war, total GDP grows at an average compounded annual rate of over 7.5 per cent (low projection) or over 9 per cent (high pro-

TABLE 8.9 (*continued*)

B. PROJECTIONS

Item	1969	1970	1971	1972	1973
LOW					
1. Value added to GDP:					
Calculated as a residual:					
(15) Proceeds from export sales	133	266	313	382	451
(16) Proceeds from local sales	1	9	12	13	14
(17) Net foreign capital flow	65	40	38	35	33
(18) Imports of materials	23	25	23	22	22
(19) Imports of services	46	50	46	44	44
(20) Local goods and services	26	30	32	32	32
(21) Value added to GDP^b (15)+(16)+(17)− (18)−(19)−(20)	104	210	262	332	400
Calculated from individual components:					
(22) Payments to government	23	68	116	144	176
(23) Harbor dues and port charges	3	7	9	11	13
(24) Local wages and salaries	4	4	4	4	4
(25) Net factor income paid abroad	74	131	133	173	207
(26) Value added to GDP^b (22)+(23)+(24)+(25)	104	210	262	332	400
2. Value added to GNP:					
Calculated as a residual:					
(27) Value added to GNP^b (15)+(16)+(17)−(18)− (19)−(20)−(25)	30	79	129	159	193
Calculated from individual components:					
(28) Value added to GNP^b (22)+(23)+(24)	30	79	129	159	193

(*Table continued on p. 119*)

jection), principally on the basis of increments in value added from petroleum. Petroleum alone is responsible for causing total GDP to grow at nearly 4.5 per cent (low projection) or nearly 6 per cent (high projection).

It has been stressed in an earlier discussion that national rather than domestic accounting concepts are more relevant in dealing with economies possessing large foreign-owned extractive export sectors (see Chapter Five and Appendix E). National accounting concepts are thus employed in Table 8.11. The annual GNP figures of Table 8.11 are smaller than their counterparts in Table 8.10 by

TABLE 8.9 (*continued*)

B. PROJECTIONS (*continued*)

Item	1969	1970	1971	1972	1973
HIGH					
1. Value added to GDP:					
Calculated as a residual:					
(29) Proceeds from export sales	133	304	394	500	602
(30) Proceeds from local sales	1	9	12	13	14
(31) Net foreign capital flow	65	49	46	44	42
(32) Imports of materials	23	27	29	27	27
(33) Imports of services	46	54	58	53	53
(34) Local goods and services	26	35	40	40	40
(35) Value added to GDP[b] (29)+(30)+(31)— (32)—(33)—(34)	104	246	325	437	538
Calculated from individual components:					
(36) Payments to government	23	80	144	186	240
(37) Harbor dues and port charges	3	9	11	14	17
(38) Local wages and salaries	4	4	4	4	4
(39) Net factor income paid abroad	74	154	165	232	276
(40) Value added to GDP[b] (36)+(37)+(38)+(39)	104	246	325	437	538
2. Value added to GNP:					
Calculated as a residual:					
(41) Value added to GNP[b] (29)+(30)+(31)—(32)— (33)—(34)—(39)	30	92	160	205	262
Calculated from individual components:					
(42) Value added to GNP[b] (36)+(37)+(38)	30	92	160	205	262

Data from Tables 5.2, 8.1, and 8.2, and author's estimates.
[a] Dash indicates less than £500,000 Nigerian.
[b] May not equal sum of components because of rounding.

the annual amounts of net factor payments made abroad. As expected, the compound rates of growth of GNP (over 6 per cent for the low projection and over 7 per cent for the high projection) are lower than those of GDP (over 7.5 per cent and over 9 per cent, respectively) for the period 1969–73. The oil sector contribution is much less to growth of GNP than to growth of GDP. If the non-petroleum sectors stagnated at 1969 levels, petroleum value added to GNP alone would cause total GNP to increase at slightly more than 2.5 per cent (low projection) or over 3.5 per cent (high pro-

TABLE 8.10. CONTRIBUTION OF PETROLEUM VALUE ADDED TO
NIGERIAN GROSS DOMESTIC PRODUCT, 1963–73
(Million £ Nigerian, except as otherwise indicated)

A. HISTORICAL

Value added	1963	1964	1965	1966	1967	1968
Petroleum value added	8	17	32	44	51	18
Nonpetroleum value added	1,303	1,308	1,381	1,419	1,434	1,463
Gross Domestic Product (market prices)	1,311	1,325	1,413	1,463	1,485	1,481
Petroleum value added as per cent of GDP	1	1	2	3	3	1

B. PROJECTIONS

Value added	1969	1970	1971	1972	1973
		LOW			
Petroleum value added	104	210	262	332	400
Nonpetroleum value added[a]	1,495	1,555	1,617	1,682	1,749
Gross Domestic Product (market prices)	1,599	1,765	1,879	2,014	2,149
Petroleum value added as per cent of GDP	7	12	14	16	19
		HIGH			
Petroleum value added	104	246	325	437	538
Nonpetroleum value added[a]	1,495	1,555	1,617	1,682	1,749
Gross Domestic Product (market prices)	1,599	1,801	1,942	2,119	2,287
Petroleum value added as per cent of GDP	7	14	17	21	24

Data are from Tables 5.3, A.1, and author's estimates. All figures pertain to domestic product national income accounting concepts.

[a] The 1969 figure is estimated on the basis of a 2.2 per cent growth rate; 1970–73 figures assume a 4.0 per cent annual growth rate.

jection). But no matter how one measures it, the direct contribution of the petroleum industry to Nigerian value added should be very large and increasing.

Application of the Requirements Analysis

This chapter has dealt so far with the likely extent of the Nigerian petroleum industry's contributions of three critical factors of production—foreign exchange, investment resources, and skilled labor. Information used in making these factor projections

TABLE 8.11. THE CONTRIBUTION OF PETROLEUM VALUE ADDED TO
NIGERIAN GROSS NATIONAL PRODUCT, 1963–73
(*Million £ Nigerian, except as otherwise indicated*)

A. HISTORICAL

Value added	1963	1964	1965	1966	1967	1968
Petroleum value added	8	16	18	24	32	18
Nonpetroleum value added[a]	1,286	1,284	1,366	1,401	1,413	1,443
Gross National Product (market prices)	1,294	1,300	1,384	1,425	1,445	1,461
Petroleum value added as per cent of GNP	1	1	1	2	2	1

B. PROJECTIONS

Value added	1969	1970	1971	1972	1973
LOW					
Petroleum value added	30	79	129	159	193
Nonpetroleum value added[a]	1,475	1,534	1,595	1,659	1,725
Gross National Product (market prices)	1,505	1,613	1,724	1,818	1,918
Petroleum value added as per cent of GNP	2	5	7	9	10
HIGH					
Petroleum value added	30	92	160	205	262
Nonpetroleum value added[a]	1,475	1,534	1,595	1,659	1,725
Gross National Product (market prices)	1,505	1,626	1,755	1,864	1,987
Petroleum value added as per cent of GNP	2	6	9	11	13

Data are from Tables 5.3, A.1, and author's estimates. All figures pertain to national product national income accounting concepts.

[a] The 1969 figure is estimated on the basis of a 2.2 per cent growth rate; 1970–73 figures assume a 4.0 per cent annual growth rate.

has also been drawn on to project the petroleum sector's contributions to Nigerian GDP and to GNP. Non-oil value added for both concepts has been projected arbitrarily to provide bases for comparison. But to this point the most interesting questions concerning factor contributions have not yet been posed. Large and growing contributions of foreign exchange and investment resources by the petroleum industry have a real welfare impact on the Nigerian economy only to the extent that they can successfully be translated into increases in Nigerian national income in the non-

petroleum economy. Clearly the addition of factors so readily available that their productive employment is precluded will do nothing toward increasing total output. It is therefore of interest to construct an input-output model of the Nigerian economy and to employ it in a requirements analysis in order to identify which factor (or factors) of production is constraining further increases in output during each year of the projection period and to determine total and sectoral gross output, deliveries to final demand, and value added for each year. It is then possible to calculate the shadow prices of the binding factors.

It will be recalled that the basic equation of the model, derived in Chapter Four (pp. 50–54) is as follows:

$$T = F \cdot X, \text{ or}$$
$$T = F \cdot (I - A)^{-1} \cdot Y.$$

The U vector of factor availabilities, with dimensions $m \times 1$, does not enter the basic equation directly. It is projected separately and then serves as a basis of comparison for the calculated T vectors.

The following variables are chosen for the application of this model to Nigerian petroleum for the period 1969–73. The economy is divided into four sectors: petroleum; agriculture, including fishing and forestry; manufacturing and construction plus nonpetroleum mining; and services, a residual category that comprises public utilities, transport, trade, distribution, and other items not included elsewhere. In terms of the model notation, n equals 4 in the A and F matrices and the X and Y vectors. Three factors of production are considered—foreign exchange, investment resources, and skilled labor. Returning to the notation of the model, m equals 3 in the F matrix and in the T and U vectors. High projection and low projection impacts of the petroleum sector are compared in separate applications of the model.

The petroleum sector receives special treatment in the construction of the model. It is included as a separate sector to emphasize the fact that there are significant intersectoral flows from petroleum to the rest of the economy and from the other sectors to petroleum. In light of the oil industry's ample sources of the three

factors considered, the model is formulated so as to allow the petroleum sector to require and obtain precisely the amounts of factors needed to produce exogenously projected levels of factor contributions, value added to GDP and deliveries to final demand. In order to accommodate the petroleum sector in this fashion, separate petroleum columns are included in the F and A matrices for each year of each projection. This procedure causes all elements of the $(I - A)^{-1}$ matrix to alter from year to year. In addition, petroleum is treated separately with regard to factor contributions. All contributions of factors by the petroleum sector are projected exogenously. Each annual level is the sum of the sector's own needs (assumed to be supplied internally) plus its net contribution to other sectors. To gauge how much growth can be attributed to the existence of petroleum, one must know separately the levels of each factor contributed by petroleum and by the other sectors.

A few additional assumptions are necessary before the model can be run. The petroleum sector's annual levels of GDP for all years of the projections are as given in Table 8.9. GDP rather than the more appropriate GNP is employed in the model for comparability in light of the fact that Nigeria officially employs the domestic income accounting concept. In the presentation of the results of the model, qualifications are made to allow GNP accounting and consideration of transfer prices rather than posted prices as the basis for calculating value added to GDP. Initial target growth rates of GDP for the non-oil sectors are as follows: agriculture, 2.8 per cent; manufacturing and construction, 8.4 per cent; and services, 6.0 per cent.[3] Each sector's value added differs from its gross output by the amount of its intersectoral purchases. Since these are assumed to take place in fixed proportions, the assumptions about growth rates of value added are also assumptions about growth rates of gross output.[4] Although the final growth rates achieved are determined by the model, this assumption sets the initial allocation of total value added among the three non-oil sectors.

The availability of investment resources in each year is calcu-

lated on the assumption that nonpetroleum domestic plus foreign investment in that year is equal to 12 per cent of nonpetroleum GDP in the preceding year.* Stated differently, it is assumed that 12 per cent of GDP accruing from all sectors apart from oil in any given year is invested in the non-oil economy during the following year. This percentage is the average relationship between gross fixed capital formation and GDP during the period 1962–66.

Any results obtained by using the model are only as reliable as the data employed in it. Complete details describing the assumptions that underlie the projected data are contained in Appendix D. The petroleum sector's factor contributions can be dispatched quickly. Earlier sections of this chapter provide the basis for the contributions of petroleum to the projected U vectors of factor availabilities. Table 8.8 contains the oil balance-of-payments impacts. For each year the total of this figure plus the petroleum industry's own requirements of foreign exchange is the sector's contribution of foreign exchange. The petroleum sector's portion of the second element of the U vector, investment resources, is found by adding the industry's payments to government in Table 8.3 to a projection of the industry's own investment. This procedure implicitly assumes that there is no substitution of government revenues associated with petroleum for other revenues and that nonpetroleum revenues grow equally as fast as government recurrent expenditures. The sensitivity of the results to changes in this arbitrary assumption is tested indirectly below by reducing in steps the postulated relationship between the availability of nonpetroleum investment resources in one year and nonpetroleum GDP in the prior year. Finally, the petroleum sector's contribution of skilled labor, assumed to approximate its own needs, is estimated as two-thirds of the projections of local wages and salaries in Table 8.2.

Projections of the nonpetroleum sectors' contributions of factors require more space to describe. The relevant appendix tables in this regard are Tables D.6 and D.8. The projections of for-

---

* This assumption is fragile not only because of the fairly arbitrary choice of an investment percentage but also because of the implicit understanding that the total available investment resources can be allocated among sectors as required.

eign exchange availability for the non-oil Nigerian economy are taken directly from Table 8.8 and are based on a detailed forecast of the entire Nigerian balance of payments that assumes no changes in the level of foreign exchange reserves. To repeat, the generation of investment resources apart from petroleum is postulated to occur such that the amount available in each year equals 12 per cent of the previous year's nonpetroleum GDP. Note that this assumption is not a domestic saving rate, in view of the fact that foreign (nonpetroleum) capital inflows as well as private and public domestic saving are included in investment resources. The most elusive factor contributions are the projected supplies of skilled labor. This tenuous projection is carried out on the basis of fragmented bits of historical information, as reference to the notes of Table D.6 shows. In short, from an estimated total of 332,000 skilled laborers in Nigeria in 1968, the availability of this factor is postulated to grow at an annual rate of 5.75 per cent. For convenience the number of skilled laborers is converted to a financial basis with the assumption of an average salary rate of £360 per man.

Along with the separate projections of the U vectors, there are two other inputs of data required to employ the model—the Leontief inverse matrices $[(I-A)^{-1}]$ and the F matrices of coefficients of factor requirements. The A matrices of input-output coefficients and the derived inverse matrices are updated versions of Carter's pioneering study.[5] Appendix D contains A and $(I-A)^{-1}$ matrices for the four historical years 1965–68. For the five years of the projection, 1969–73, there are separate sets of matrices based on first the high and then the low projections. In each year separate entries are included for the columns pertaining to the petroleum sector. The remaining columns in each A matrix are calculated for 1965 and are assumed to remain fixed. Naturally all elements of the inverse matrices alter slightly when the petroleum elements are changed.

The F matrices of factor requirement coefficients for 1965–73 are also included in Appendix D. These three-by-four matrices contain separate petroleum columns for each year. The model is constructed so that the petroleum sector supplies itself with ex-

actly the amounts of factors that a priori have been established as its needs. Separate F matrices are thus constructed for the high and low projections. Initially F matrices were calculated for 1965 and 1966. (The years 1967 and 1968 were excluded from consideration because of civil war influences.) The nonpetroleum entries in the F matrices for 1969–73 are based on the corresponding elements for these two historical years. The foreign exchange row is an average of the two years, and the 1966 coefficients are used in the rows for investment resources and skilled labor. In estimating the foreign exchange rows, all consumer imports are distributed among the four sectors of the model on the basis of GDP shares, reflecting the economy's need (perhaps politically based) for a minimum level of such imports. Sectoral intermediate imports are adjusted aggregates from Carter's input-output study. The non-oil requirements of investment resources are achieved by adjusting Lewis's 1963 breakdown of gross fixed capital formation.[6] The skilled labor coefficients are put together using information on numbers of skilled labor employed per sector contained in a recent study of Nigerian human resources.[7]

Table 8.12 summarizes the results obtained from applying the

TABLE 8.12. INPUT-OUTPUT MODEL: SUMMARY OF RESULTS USING HIGH PROJECTION
(*Million £ Nigerian*)

Projection	1969	1970	1971	1972	1973
		DATA INPUTS			
Factor availabilities:					
Foreign exchange:					
Petroleum	198	353	440	544	644
Nonpetroleum	240	200[a]	265	275	285
Total	438	553[a]	705	829	929
Investment resources:					
Petroleum	108	195	274	291	345
Nonpetroleum	176	177	188	199	207
Total	284	372	462	490	552
Skilled labor:					
Petroleum	3	3	3	3	3
Nonpetroleum	127	135	143	151	159
Total	130	138	146	154	162

TABLE 8.12 (continued)

Projection	1969	1970	1971	1972	1973
	RESULTS OF THE MODEL				
Factor uses:					
Foreign exchange:					
Petroleum	143	235	252	312	356
Nonpetroleum	295	318	338	355	373
Total	438	553	590	667	729
Investment resources:					
Petroleum	85	115	130	105	105
Nonpetroleum	196	213	229	242	257
Total	281	328	359	347	362
Skilled labor:					
Petroleum	3	3	3	3	3
Nonpetroleum	123	134	143	151	159
Total	126	137	146	154	162
Gross output:					
Petroleum	124	281	367	478	579
Nonpetroleum:					
Agriculture	823	861	893	911	926
Mfg. and construction	477	526	567	610	653
Services	524	565	605	636	666
Total	1,824	1,952	2,065	2,157	2,245
Total gross output	1,948	2,233	2,432	2,635	2,824
Gross Domestic Product:					
Petroleum	104	246	325	437	538
Nonpetroleum:					
Agriculture	820	858	890	908	923
Mfg. and construction	202	223	244	262	281
Services	455	491	525	552	579
Total[b]	1,476	1,571	1,658	1,722	1,782
Total GDP	1,580	1,817	1,983	2,159	2,320
Deliveries to final demand:					
Petroleum	113	269	353	464	564
Nonpetroleum:					
Agriculture	717	745	768	776	782
Mfg. and construction	350	378	407	441	476
Services	403	428	455	477	499
Total	1,470	1,551	1,630	1,694	1,757
Total deliveries to final demand[a]	1,583	1,820	1,983	2,158	2,321

Data from Appendix Table D.43 and author's estimates.

[a] Entry is based on an assumption that a backlog of £ 50 million of nonpaid imports is eliminated.
[b] Sum of individual components may not add to total because of rounding.
[c] May not equal total Gross Domestic Product because of rounding.

input-output model with high projection assumptions. Foreign exchange is found to be the binding constraint during 1969 and 1970. In each of these years one additional unit of foreign exchange would allow the economy to produce 4.98 incremental units of non-oil GDP. During these two years, therefore, the shadow price of foreign exchange contributed by petroleum is about five units of non-oil value added for the Nigerian economy. In this respect it should be noted that the results for 1970 summarized in Table 8.12 depend critically on an assumption regarding the schedule for future repayment of a £50 million backlog of nonpaid imports. If Nigerian officials could postpone repayment until foreign exchange is no longer scarce, then more growth is possible according to the assumptions underlying the model.

In the three later years of the projection, skilled labor emerges as the binding factor constraint. An extra unit of skilled labor in 1971 would result in 11.54 additional units of non-oil GDP. The shadow price for 1972 is 11.45 and for 1973 it is 11.33.* Beginning in 1971, therefore, Nigerian growth should be constrained by a scarcity of skilled labor, a factor that the petroleum industry supplies in minimal amounts. One would expect that Nigerian growth, fueled by rapidly increasing contributions of foreign exchange and domestic saving from petroleum, would eventually run up against an absorptive capacity limit. But it is especially interesting that this might in fact occur as early as 1971.

The model also yields sectoral deliveries to final demand, gross output, and GDP, as summarized in Table 8.12† It is instructive to compare the totals for non-oil GDP with those projected arbitrarily in the preceding section of this chapter, where the assumed rates were 2.2 per cent during 1969 and 4.0 per cent thereafter.

---

* These high shadow prices would presumably be altered to more realistic levels if the model allowed some substitution among factors, e.g. using foreign exchange to purchase imported skilled labor.

† Since imports are treated outside of the input-output matrix, total deliveries to final demand equal total value added (GDP). The relation between sectoral gross output and value added, stated in note 4, p. 216, is employed to calculate sectoral value added in this model.

Petroleum in Nigeria: Some Projections 129

The model results show these rates to be somewhat conservative. During the five years of the projection, the following rates of growth of nonpetroleum GDP are achieved (percentages beginning with 1968–69 and continuing through 1972–73): 0.9, 6.4, 5.6; 3.8, and 3.5. The compound rate between 1969 and 1973 is 4.8 per cent. The inclusion of the petroleum sector increases the annual rates to the following per cent levels: 6.7, 15.0, 9.2; 8.8, and 7.5. Over the five years of the projection the compound rate of growth of total GDP is 10.1 per cent. Excluding the war year 1969, the compound rate over the final four years is 8.5 per cent.

At first glance, the outlook for medium-term growth of the Nigerian economy appears extremely optimistic. The growth rates for the nonpetroleum economy are high but attainable. Several adjustments, however, must be made to the contributions of the petroleum sector to Nigerian value added. For reference, on the basis of the high projection, petroleum value added to GDP is the following for the projection period 1969–73, respectively (million £ Nigerian): 104, 246, 325, 437, and 538. It has already been argued that GNP accounting is more appropriate than GDP accounting for economies with foreign-owned export sectors. In contrast to the very high oil contributions to GDP, comparable figures for petroleum value added to GNP are (in million £ Nigerian): 30, 92, 160, 205, and 262 (see Table 8.9, p. 119). The compound growth rate of total GNP is nearly 8 per cent between 1969 and 1973 and about 7 per cent between 1970 and 1973.

A second adjustment concerns the evaluation of crude petroleum at realized prices rather than at posted prices. The actual prices at which the individual petroleum corporations transfer crude production from Nigeria for refining elsewhere are not commonly known outside the industry. For sake of argument it can be assumed arbitrarily that the average realized price for the industry will be \$1.85 per barrel of crude exported. The petroleum industry's value added to GDP then reduces to the following for 1969–73, respectively (million £ Nigerian): 94, 222, 299, 406, and 501. Note that this adjustment is necessary only if GDP ac-

counting is employed. In the calculation of petroleum value added to Nigerian GNP, it makes no difference how one evaluates the industry's total proceeds (see Appendix E).

It is useful to examine the sensitivity of these results with respect to changes in some of the key assumptions employed. Certainly a major assumption concerns the levels of projected petroleum output. For comparative purposes, results of applying the model with the low projection are displayed in Table 8.13. Because the assumption of skilled labor availability is identical to that used in Table 8.12, the low projection results differ only during 1969 and 1970. During these years, a lower availability of foreign exchange causes less growth of output than with the high projection. Returning to the high projection, investment resources were not found to be the binding constraint in any of the five years. But if one alters the assumed investment percentage from 12 per cent to 11 per cent of the previous year's non-oil GDP, investment resources bind in 1969. Lowering the assumed percentage to 10 per cent causes this factor to constrain in 1970. Thereafter, however, the rate must be dropped to 7 per cent before investment resources become scarce.

Changes have been made in the non-oil elements of the matrix

TABLE 8.13. INPUT-OUTPUT MODEL: SUMMARY OF
RESULTS USING LOW PROJECTION
(*Million £ Nigerian*)

Projection	1969	1970	1971	1972	1973
		DATA INPUTS			
Factor availabilities:					
Foreign exchange:					
Petroleum	198	353	351	417	484
Nonpetroleum	240	200[a]	265	275	285
Total	438	553[a]	616	692	769
Investment resources:					
Petroleum	103	173	226	244	276
Nonpetroleum	176	163	169	189	200
Total	279	351	405	444	483
Skilled labor:					
Petroleum	3	3	3	3	3
Nonpetroleum	127	135	143	151	159
Total	130	138	146	154	162

Petroleum in Nigeria: Some Projections 131

TABLE 8.13 (*continued*)

Projection	1969	1970	1971	1972	1973
	RESULTS OF THE MODEL				
Factor uses:					
Foreign exchange:					
Petroleum	143	206	202	239	273
Nonpetroleum	295	300	339	357	370
Total	438	506	541	596	643
Investment resources:					
Petroleum	80	105	110	100	100
Nonpetroleum	196	201	230	244	258
Total	276	306	340	344	358
Skilled labor:					
Petroleum	3	3	3	3	3
Nonpetroleum	123	126	143	151	159
Total	126	129	146	154	162
Gross output:					
Petroleum	124	240	294	364	432
Nonpetroleum:					
Agriculture	823	813	890	908	922
Mfg. and construction	477	496	574	617	661
Services	524	533	603	634	664
Total	1,824	1,842	2,067	2,159	2,247
Total gross output	1,948	2,082	2,361	2,523	2,679
Gross Domestic Product:					
Petroleum	104	210	262	332	400
Nonpetroleum:					
Agriculture	820	810	887	905	919
Mfg. and construction	202	210	243	261	280
Services	455	463	523	550	576
Total[b]	1,476	1,483	1,653	1,716	1,775
Total GDP	1,580	1,693	1,915	2,048	2,175
Deliveries to final demand:					
Petroleum	113	228	281	350	417
Nonpetroleum:					
Agriculture	717	703	764	772	777
Mfg. and construction	350	359	419	453	488
Services	403	406	456	478	499
Total	1,470	1,468	1,639	1,703	1,764
Total deliveries to final demand[c]	1,583	1,696	1,920	2,053	2,181

Data from Appendix Table D.43 and author's estimates.
[a] Entry is based on an assumption that a backlog of £50 million of nonpaid imports is eliminated.
[b] Sum of individual components may not add to total because of rounding.
[c] May not equal total Gross Domestic Product because of rounding.

of factor requirements to allow for possible changes in factor combinations employed, a not unlikely occurrence in the event of otherwise surplus factors. One might postulate that Nigerian production processes would become more import-intensive if foreign exchange were slack. When all non-oil elements of the foreign exchange row of the F matrix are increased by 33 per cent beginning in 1970, foreign exchange replaces skilled labor as the binding constraint. Or one could assume that more capital-intensive production processes would be employed if investment resources were available in excess supply. An increase of 2 per cent of the non-oil elements in the domestic saving row of the F matrix causes investment resources to become the constraining factor in 1969, but the post-1969 results are insensitive to increases of 20 per cent. Finally, enlarging the non-oil elements of the skilled labor row in the F matrix by 3 per cent in 1969 and 1 per cent in 1970 causes skilled labor to be binding in these years. Naturally one could undertake the same type of analysis with the elements of the Leontief inverse matrix. After 1970, the results of the model are robust in the sense that they are relatively insensitive to changes in the principal assumptions employed. Hence it can be predicted with considerable confidence that skilled labor will replace foreign exchange as the factor constraining additional Nigerian growth in the very near future.

Linkage Effects

Speculation about future linkage effects is a difficult undertaking. Some possible future opportunities for the emergence of important intersectoral effects associated with the production of crude oil in Nigeria, however, can be sketched broadly. Forward linkages are examined at somewhat greater length in view of the reasonable expectation of their future significance.

The potential for important backward linkages developing vis-à-vis the Nigerian petroleum industry is not very encouraging. Evidence of this can be gleaned from examination of the petroleum columns of the Leontief inverse matrices in Appendix D to find intermediate demands by the petroleum industry from the other

Petroleum in Nigeria: Some Projections

sectors per unit of petroleum produced. For each year multiplication of the petroleum row of the inverse matrix by the vector of final demand gives the petroleum sector's gross output. The difference in each year between the petroleum sector's gross output and its value added equals its total use of purchased inputs. From this total one should subtract the inputs that the petroleum sector purchases from itself to arrive at the total amounts of inputs purchased from other industries, the relevant figure for study of backward linkages.[8] This relationship can be quantified for both the high and low projections by referring to the projections of the petroleum sector's value added and gross output in Tables 8.12 and 8.13 and to the relevant A matrixes in Appendix D. In no instance are the agricultural, manufacturing, or services sectors' inputs into the petroleum sector very large. Accordingly the possibility of achieving economies of scale, externalities, or use of underused resources is fairly small. For example, using high projection data, the petroleum sector's intersectoral purchases level off at about £40 million after 1970.

An earlier discussion indicated why the exploring and producing companies' purchases of local goods and services are not likely to expand much. There is scope for Nigerian entrepreneurs to participate more widely in servicing and supplying activities. Many of the goods that the petroleum industry now imports could and presumably will be produced in Nigeria. But not too much of this is expected to happen in the period considered here. There is much less chance that Nigerian factors will be able to substitute for the highly technical imported services of the oil industry. Until Nigerian industrialization and technical expertise expand considerably, backward linkages with the petroleum sector are likely to be limited.

A large portion of oil sector inputs into other Nigerian sectors consists of sales to the refinery. The output of Nigeria's refinery is expected to grow in step with domestic demand until about 1975. The prospect of an additional refinery is being discussed with increasing regularity. But unless inroads can somehow be made on the export market, construction of new refineries in Nigeria would

seem to be based more on political than on economic considerations (see pp. 92–95).

Natural gas has a near-zero opportunity cost to the Nigerian economy. But Nigerian utilities and industries have barely begun to tap the extremely wide potential for energy based on natural gas. It is indeed regrettable that the true extent of Nigerian natural gas resources was either not known or disregarded when the decision was made to construct the Kainji Dam for hydropower.[9] Some observers now fear that the government will institute a policy forcing the use of Kainji electricity at a price higher than that at which gas-based electricity could be produced and thereby discourage the establishment of energy-intensive industries in Nigeria. Wider use of natural gas could also be promoted through lowering the price (recently 20 pence per thousand cubic feet) set by the producing companies. The relatively high price is partially explained by the following: "Natural gas is regarded by oilmen as something of a nuisance; it is an unasked for byproduct of their activities and for them its profitability is relatively low. Often they would prefer to burn it all off and are only prevented from doing so by the outcry of the host government against such a waste of natural resources."[10]

In 1966, Nigeria narrowly missed out on an excellent opportunity to export liquefied natural gas to the United Kingdom. Conch Methane Services, a corporation formed by Shell, Continental Oil, and Union Stockyard and Transit Company of Chicago, had proposed to the UK Gas Council that it supply 10 per cent of British gas requirements from Nigeria. While the Nigerians and the British dickered over the price, natural gas was discovered in the North Sea. Notwithstanding this development, the British have since concluded deals with both Algeria and Libya to supply natural gas to Britain. But discussions with Nigeria have remained suspended.

In addition to expanded use of natural gas for energy purposes, two potential forward-linked industries that use natural gas as a raw material in their production processes might well be estab-

lished in Nigeria.[11] Recently, petrochemical and chemical industries were placed second only to agriculturally based industries in a list of industrial priorities prepared by the Federal Ministry of Industries.[12] Investigations are now under way toward the creation of a chemical complex that would produce polyvinyl chloride (for plastic shoes, raincoats, etc.), polyethylene (for plastic packaging materials, etc.), and caustic soda (for soap, etc.). This complex would be an import-substitution industry with total sales of about £6 million annually and a significant net foreign exchange saving. The economic viability of this project depends very much on the price charged for natural gas, not only because of the role of gas as raw material input but also because of the large amounts of electricity consumed in the production process (assuming that natural gas-based electricity and not Kainji electricity is to be used). This project could be implemented within the period considered here.

A second industry using natural gas as a raw material that might feasibly be established in Nigeria is a nitrogenous fertilizer plant. The major constraints on the economic viability of this venture are the limitation of the Nigerian market for nitrogenous fertilizers and the difficulty of exporting in light of recent trends toward self-sufficiency in virtually all of the major world markets. Studies indicate that a small, 100 ton/day ammonium plant would be economical in Nigeria if the domestic demand for calcium ammonium nitrate or ammonium sulfate fertilizers were to reach 100,000 tons per year.[13] Nigerian consumption of nitrogenous fertilizers has grown from 3,000 tons in 1963 to 29,000 tons in 1967. Conservative projections see this consumption figure doubling by 1973. But others predict that fertilizer consumption might grow even more rapidly in Nigeria, emulating the recent experience of several other developing countries.[14] If consumption shows continued large increases or if the plant can be made to run profitably at lower output levels, then Nigeria would gain an industry that produced output valued at about £1 million per year, with further effects of great potential importance in the Nigerian agri-

cultural sector.* In addition to this relatively small fertilizer plant, some thinking has gone into the possibility of setting up a large urea complex for export. For the time being, this seems to be ruled out because the wellhead-delivered price of natural gas, as established by the petroleum producing companies, is too high for the plant to be economical.[15]

Much of what has been said about backward linkages applies equally well to final demand linkages. The amounts of factor payments in Nigerian currency that will remain in the country to have an impact are severely restricted by the small range of input mixes likely to be employed in producing petroleum. Locally paid wages and salaries associated with the petroleum exploring and producing industry and its suppliers and contractors might grow to £10 million. Domestically expended residual payments (profits, interest, depreciation, etc.) are restricted by the high degree of expatriate ownership both in the petroleum industry and among its suppliers and contractors. In light of this, final demand linkages are likely to be very limited, though not wholly insignificant. The two technological linkages—labor and management training, and construction and maintenance of roads—that were found to be important in Chapter Seven should continue, but in neither case is there latitude for much expansion of benefits. Finally, the evolution of fiscal linkages directly involves subsequent government policy decisions, a topic reserved for the concluding chapter.

* It should be noted that the contribution of fertilizers is enhanced if their application is accompanied by greater concern over plant population, increased use of insecticides, and introduction of new and improved varieties of seeds. In discussing the potential importance of fertilizers in Nigerian agricultural development, Bruce F. Johnston, in "Agriculture's Role in Nigerian Development Strategy" (U.S. Agency for International Development, Washington, D.C., 1966), pp. 34–36, points out that fertilizers not only increase output, but also are labor-using and neutral with respect to scale.

CHAPTER NINE

The Politics of Nigerian Oil

The analysis in this study is primarily economic in nature. But petroleum, in Nigeria as elsewhere, has necessarily had important political ramifications. Indeed, some of these political effects have to be brought to the surface before one can usefully discuss the implications of this economic analysis of petroleum in Nigeria for future policy decisions.[1]

For several years after its initial discovery, petroleum remained in the background of the Nigerian political scene. This was mainly a reflection of the limited knowledge of the true extent of Nigerian reserves as well as unfamiliarity with the ramifications that oil production might have. Though speculation about oil occurred before 1964, it was only about this time that rumors of the potential importance of petroleum began to enter importantly into political discussions.[2] Any relative disinterest in petroleum that remained in 1965 came to an abrupt halt when the Federal Prime Minister in a statement to the Chamber of Commerce and the Federal Minister of Finance in his annual budget address spoke optimistically about the balance-of-payments impacts that oil production would have in Nigeria. Political feelings about petroleum changed from apathy to euphoria. Interest in controlling the newly recognized benefits from oil grew apace.

The emergence among Nigerian leaders of widespread knowledge about the potential benefits that the production of petroleum might bestow on Nigeria coincided with the eruption of another in the series of recurrent political crises that beset Nigeria periodically from independence in October 1960 to the secession of Bi-

afra in May 1967. The military coup of January 1966, engineered and led primarily by officers of the Ibo ethnic group (mainly of Eastern Nigeria), shattered the fragile coalition between Northern- and Eastern-led political parties that had dominated postcolonial federal politics. Following this first coup, threats of secession were heard from disgruntled Northerners. But in July 1966 a second military coup took place, this one led by Northern officers. Although these Northern officers in many instances were themselves representatives of minority ethnic groups in the North, initially it was expected that they would take the North out of the Nigerian Federation. Rather than do this, the new military government promised to reintegrate the country as a federation of states and in the process to reduce the former predominance of the North in Nigerian federal politics.*

In response, threats of secession began to come from the East, especially in face of the periodic massacres of Ibo tribesmen living in the North that occurred before, during, and after the second coup. At first glance it seemed curious that the Northerners would not allow the Ibos to leave the Federation and form a separate nation. Yet there were strong arguments against this line of action. Clearly chauvinistic desires as well as solid economic reasons argued in favor of maintaining Africa's most populous country as a single unit. In addition, secession of what was then the Eastern Region would have placed several minority tribal groups, numbering perhaps four million people, under the domination of the Ibo people, whose population in the East at that time probably approached seven million.

These arguments were firmly buttressed by the growing awareness of Nigeria's potential as a producer of crude petroleum. Approximately two-thirds of petroleum production under way in 1966 was located in the Eastern Region, most of it in areas traditionally inhabited by the minority groups, not the ruling Ibos. Moreover, legislation pertaining to the distribution of petroleum-related payments to the Nigerian governments resulted in a very significant

* In April 1968 the existent four-region federation was replaced by one consisting of twelve states.

transfer from the regions of origin to the federal government and to the nonproducing regions. The existence of petroleum in Nigeria was probably not responsible for the schism that ultimately erupted into civil war in July 1967. Integral and paramount factors included the long-standing political rivalry, mutual distrust, and in some instances deep hatred among the three largest tribal groupings of sub-Saharan Africa (Hausas, Ibos, and Yorubas); Ibo middle-level economic predominance in non-Ibo and especially Northern areas; the Northern massacres of Ibo tribesmen; and minority tribes' desires for greater political control. But oil may very well have been the extra ingredient that finally precipitated the military conflict.[3]

Implications of the Location of Petroleum-producing Areas

The history of crude oil production in Nigeria has been summarized in Chapter Five, and especially in Table 5.1. But so brief a treatment naturally allows no room for consideration of the political factors, such as where the oil production occurs and which factor owners benefit from the net gains associated with petroleum, that play a large role in determining the impact of foreign investment in petroleum on Nigerian growth prospects.

General knowledge of the existence of large reserves of petroleum has exacerbated divisive ethnic pressures. At issue is the location of ethnic groups, state boundaries, and oil-producing areas.[4] Crude petroleum has been produced in only two of Nigeria's former regions (Eastern and Mid-Western) or four of the present states (Rivers, East Central, South Eastern, and Mid-Western). The Ibo people have a clear population majority in the former Eastern Region and especially in the East Central State. But minority groups significantly outnumber the Ibos in the Rivers State, where most of the former Eastern Region oil was produced, as well as in the South Eastern State. In the former Mid-Western Region, now the Mid-Western State, there is a strong minority of Ibos, but non-Ibo groups have generally been in control—with the exception of the brief takeover of the Mid-West by Biafra in August–September 1967. If the Ibo-led secession of Biafra, the

TABLE 9.1. CRUDE PETROLEUM PRODUCTION BY FIELD AND STATE, APRIL 1967
(*Barrels per day, except as otherwise indicated*)

Field	Number of wells	Rivers State	East Central State	East Central and Rivers States	Mid-Western State
Afam	4	—	5,438	—	—
Afam Umuosi	1	—	547	—	—
Agbada	11	25,194	—	—	—
Ahia	6	12,961	—	—	—
Apara	4	1,096	—	—	—
Bomu	25	76,637	—	—	—
Ebubu	7	3,440	—	—	—
Eriemu	2	—	—	—	2,956
Imo River	30	97,195	—	—	—
Isimiri	4	—	10,570	—	—
Kokori	8	—	—	—	35,562
Korokoro	6	—	—	20,519	—
Nkali	4	—	—	11,291	—
Obagi	15	40,604	—	—	—
Obigbo North	11	38,754	—	—	—
Okan	27	—	—	—	56,367
Oloibiri	7	4,418	—	—	—
Olomoro	21	—	—	—	52,261
Oweh	4	—	—	—	14,169
Remuekpe	1	37	—	—	—
Ughelli	3	—	—	—	6,299
Umuechem	14	31,865	—	—	—
Uzere East	5	—	—	—	11,861
Uzere West	10	—	—	—	21,984
Total	230	332,201	16,555	31,810	201,459
Per cent of total	...	57.1	2.8	5.5	34.6

Data from Nigeria [20], p. 7. All fields are Shell-BP except Obagi (Safrap) and Okan (Gulf).

former Eastern Region, had been successful, then the Ibos would have gained hegemony over oil-bearing areas belonging to minority groups. On the other hand, current state boundaries create inequities for the Ibos from the viewpoint of petroleum production.

To examine this problem more closely, production figures by field and state for one of the last prewar months during which the oil industry was operating under normal conditions are set forth in Table 9.1. The production breakdown between the Mid-Western State and the two producing states of the former Eastern Region was approximately one-third to two-thirds. And nearly 90

TABLE 9.2. ETHNIC ORIGIN AND STATE LOCATION OF NIGERIAN
PETROLEUM PRODUCTION, APRIL 1967

Production	Rivers	East Central	Mid-West	Total
A. Barrels per day:				
Ibo	154,907	7,181	—	162,088
Non-Ibo	188,873	29,605	201,459	419,937
Total	343,780	36,786	201,459	582,025
B. Percentages:				
Ibo	45.0	19.5	—	27.8
Non-Ibo	55.0	80.5	100.0	72.2
Total	100.0	100.0	100.0	100.0

Data from Table 9.1; Nigeria [20], pp. 14–15; K. M. Buchanan and J. C. Pugh, *Land and People in Nigeria* (London: University of London Press, 1955), pp. 79–88; Yehuda Karmon, *A Geography of Settlement in Eastern Nigeria* (Jerusalem: Magnes Press, Hebrew University, 1966), pp. 40–41.

per cent of the Eastern production originated in the Rivers State. Location of the fields in Table 9.1 is shown on the endsheet map.

Information on ethnic origin is required before any conclusions can be reached. In Table 9.2 the production data of Table 9.1 are retabulated according to tribal areas and states. It cannot be stressed too strongly that the breakdown between Ibo and non-Ibo tribal areas is at best exceedingly fragile. The concept of what areas are or are not part of the traditional domain of a particular tribe is moot in the extreme. With this in mind, the split in Table 9.2 is carried out on the basis of pre–civil war habitation of each of the producing oil field areas contained in Table 9.1. But there is still the problem of whether certain minor tribes are to be considered separate or part of the Ibo ethnic group.

Abstracting from these limitations, the results portrayed by Table 9.2 are indeed striking. With actual production of April 1967 as a base, more than one-fourth of total Nigerian oil production occurred in Ibo-inhabited areas. By far the largest portion of this was in the principally non-Ibo Rivers State, where Ibo oil areas accounted for nearly half of total production. Strangely enough, in the predominantly Ibo East Central State, Ibo areas contained only about one-fifth of total state production, which itself amounted to only 6 per cent of total Nigerian output of crude. The fact that the Rivers State produced nearly all of the oil in the

former Eastern Region has been highly publicized.[5] But it is not widely understood that perhaps 45 per cent of Rivers State production took place in Ibo-inhabited areas. Since Biafran secession has been unsuccessful, this anomaly could underscore eventual Ibo demands for wider East Central State territory. The federal government has acknowledged that state boundary lines are negotiable. Current state boundaries follow long-standing provincial lines that were established before petroleum became an issue. Hence no purposeful gerrymandering seems to have taken place in setting up the state boundaries. But expectations regarding proved or potential petroleum fields are sure to cause tremendous problems for any future commission arbitrating disputes over the new state boundaries.

Implications of the Allocation of Petroleum-derived Revenues

Given the political climate that existed in Nigeria during the 1960's, it would have been surprising indeed if the allocation of petroleum-derived revenues had not become a major political issue. The allocation procedure was not itself an explicit bone of contention in any of the more important political struggles. This issue, of course, was only a part of a much larger concern, the distribution of aggregate tax revenues between the Nigerian federal government and the states (formerly the regions). As revenues from petroleum grew, absolutely and proportionally to total government revenues, pressures were created that increased already existent ethnic rivalries. Those groups inhabiting areas where oil was produced tended to oppose the procedures in force that redistributed oil revenues, whereas groups living in areas without petroleum generally pressed for greater redistribution.

To analyze the origin and distribution of payments made by the petroleum industry to the Nigerian governments, it is necessary to review Nigerian procedures for collecting and allocating taxes, especially as they pertain to the types of payments made by the petroleum industry. Actual payments by the industry in 1967 are included to furnish examples and to provide background for Table 9.3. Normally the federal government collects about 85 per cent of total revenues, including all of its own and two-thirds of

state governments'. Among the revenues collected by the federal government are petroleum profits tax, rentals and royalties, and customs duties. In 1967 the petroleum industry made payments to the Nigerian governments totaling £27 million. Since nearly all consisted of the items just mentioned, 97 per cent of this total was paid directly to the federal government.

According to the Nigerian constitution, the division of certain types of state-oriented revenues (including oil royalties and rentals and oil-related customs duties) is as follows: state of origin, 50 per cent; federal government, 20 per cent; and distributable pool, 30 per cent. Decree Number 15 of 1967 altered this split to 50 per cent, 15 per cent and 35 per cent, respectively.*

By constitutional agreement 50 per cent of the petroleum rentals and royalties, about £10 million in 1967, thus went directly back to the states (then regions) of origin. The primary mechanism through which the federal government redistributes a further portion of its revenue to the states is the distributable pool. With 35 per cent of mining rentals and royalties and of general import duties being paid into this pool, the associated petroleum contributions in 1967 were £7.3 million. Under the former system of regions, the distributable pool reallocated its revenues on the following basis: North, 42 per cent; East, 30 per cent; West, 20 per cent; and Mid-West, 8 per cent. With the creation of the twelve states, these proportions tentatively have been subdivided without disturbing former regional shares (e.g. the former Northern Region's 42 per cent is now divided equally among the six new Northern states).†

---

* This split was altered again by the Distributable Pool Account Decree of 1970. The new distribution of revenues accruing from petroleum royalties and rentals is as follows: state of origin, 45 per cent; federal government, 5 per cent; and distributable pool, 50 per cent.

† Decree Number 15 of 1967 specified the following percentage shares for the twelve states: Western, 18 per cent; Lagos, 2 per cent; Mid-Western, 8 per cent; North Western, 7 per cent; North Central, 7 per cent; Kano, 7 per cent; North Eastern, 7 per cent; Benue-Plateau, 7 per cent; Kwara, 7 per cent; East Central, 17.5 per cent; South Eastern, 7.5 per cent; and Rivers, 5 per cent. This distribution of petroleum-derived revenues from the distributable pool was altered in early 1970. Under the new law, half of such revenues is to be distributed equally among the twelve states and half is to be allocated to the states according to population.

TABLE 9.3. HYPOTHETICAL EXAMPLE OF ORIGIN AND DISTRIBUTION OF ACTUAL 1967 PETROLEUM INDUSTRY PAYMENTS TO THE NIGERIAN GOVERNMENT
(*Million £ Nigerian, except as otherwise indicated*)

State	Royalties	Rentals	Profits tax[a]	Customs duty	Stamp duty, other	Total Amount	Total Per cent
A. Origin of payments to government:							
Mid-Western State	5.0	1.7	1.8	.2	.2	8.9	33
Rivers State	8.9	1.7	3.3	.5	.4	14.8	55
East Central State	1.0	1.8	.4	—[b]	.1	3.3	12
Total	14.9	5.2	5.5	.7	.7	27.0	100
B. Distribution of payments to government:							
Federal government		3.0	5.5	.5	—	9.0	33
Mid-Western State	4.0	—	—[b]		.2	4.2	16
Rivers State	5.6	—	—[b]		.4	6.0	22
East Central State	2.5	—	—[b]		.1	2.6	10
South Eastern State	.8	—	—[b]		—	.8	3
Two states of former Western Region	1.4	—	—[b]		—	1.4	5
Six states of former Northern Region	2.8	—		.1	—	2.9	11
Total	14.9	5.2	5.5	.7	.7	27.0	100

Data from Table 6.2 and author's estimates.
[a] Petroleum profits tax.
[b] Less than £50,000.

Table 9.3 contains a comparison of the origin and allocation of the petroleum industry's actual payments to government in 1967 under the hypothetical assumption that the current state system existed in that year. With growing oil production the distribution percentages will shift in favor of the federal government, since petroleum profits tax, which is collected by the central government and not automatically redistributed, will assume an increasingly larger share of the industry's total payments to government. This is the result of the production companies' early use of accumulated capital allowances to reduce payments of profits taxes, all of which are retained by the center.* Therefore, unless the current system for distributing oil revenues is altered, the fiscal power of the federal government vis-à-vis the state governments should grow significantly.

* The reader is referred to Chapter Two and especially to Appendix B for details concerning the calculation of payments to the Nigerian government.

But Table 9.3 has a much more immediate meaning. The implications of the redistribution of petroleum revenues were not overlooked by Biafran leaders.[6] The three states of the former Eastern Region (Rivers, East Central, and South Eastern), which made up Biafra at the outset of the war, were responsible for 67 per cent of total oil revenues. Abstracting from the secession and the war, these states would have received 35 per cent of the total from the redistribution process. Of these amounts the Ibo-dominated East Central State's shares would have been 12 per cent and 10 per cent, respectively. In addition, the Eastern States would have received benefits from a portion of the federal government's use of its oil revenues (which were 33 per cent of the total). On a population basis, one might impute about one-fourth of federal oil revenue to the Eastern States. This would give these three states an additional 8 per cent of total oil revenues, of which perhaps half would go to the East Central State. As indicated in Table 9.2, nearly 28 per cent of recent oil output was probably produced in Ibo-inhabited areas. But under the current state system, Ibo leaders can expect to control directly or benefit from only about 14 per cent of total Nigerian oil revenues, assuming they govern only the East Central State. This is a far cry from the 67 per cent of total oil revenues that would have been controlled by an independent Biafra and from the 43 per cent that might have accrued to an Eastern Region within the former federation.

A Hypothetical Analysis of the Impact of Petroleum on a Biafran Economy

In view of the failure of Biafra's bid for independence, speculating about the effects that production of petroleum might have had on the economy of an independent Biafra offers insight into the economic motivations of both sides during the Nigerian civil war. In the following analysis it is assumed that Biafra's borders are those of the former Eastern Region of Nigeria. At first glance it might seem that this assumption would ease the task of gathering and projecting economic statistics. Although this is true to a degree, the Federation of Nigeria has not recently estimated na-

tional income accounts or balance-of-payments entries on a regional basis, nor have students of the Nigerian economy come up with anything more than educated guesses about such variables. Moreover, the extent of war damage in Biafra was considerable and is difficult to estimate. This discussion should serve as fair warning concerning the fragile nature of the projections of the Biafran economy employed here. In spite of the severe shortcomings surrounding the Biafran non-oil projections, these numbers are close enough orders of magnitude to allow comparisons with the possible petroleum impacts in Biafra, their sole function in this study. It must be underscored that this brief speculation about a Biafran economy is in no way meant to be a balanced appraisal of the economic benefits and costs for Biafra in breaking away from Nigeria.

This entire section is predicated on the hypothetical assumption that the existent pattern of concessions and taxation arrangements would not have been altered in an independent Biafra. The impacts of petroleum on the Biafran economy are summarized in five tables, Tables 9.4 through 9.8, whose notes contain details of the assumptions employed. No explicit attempt is made to consider the destructive impacts of the civil war, though they clearly would have had widespread influence on the future of a Biafran economy. In general the approach used in estimating revenues, foreign exchange, and value added of the petroleum sector in Biafra is as follows. First, in Table 9.4 the low and high projections for petroleum production from Chapter Eight are divided between Biafra and the Mid-Western State of Nigeria—the only non-Biafran area in Nigeria assumed to produce oil—on the basis of oil company plans for expansion. This presents no problem, since the original output projections are built up on a company-by-company and state-by-state basis. Next the annual Biafran proportions of the united Nigeria totals are calculated. These are then applied to the petroleum variables to find the respective Biafran shares.

The division of exports in Table 9.4 indicates a rapid acceleration of production in Biafra. The Biafran share of the expanding total of Nigerian oil exports rises from less than one-third in 1969 to over one-half by 1973. As indicated in Table 9.5, petroleum

TABLE 9.4. THE DIVISION OF PETROLEUM PRODUCTION BETWEEN
BIAFRA AND MID-WESTERN NIGERIA, 1969–73
(*Volume in thousands of barrels per day, except as otherwise indicated*)

Geographical area	1969	1970	1971	1972	1973
		LOW PROJECTION			
Biafra	170	475	575	750	925
Mid-Western State	370	575	675	775	875
Total	540	1,050	1,250	1,525	1,800
Biafra as a proportion of total	.314	.452	.460	.492	.514
		HIGH PROJECTION			
Biafra	170	550	785	1,070	1,300
Mid-Western State	370	650	790	930	1,100
Total	540	1,200	1,575	2,000	2,400
Biafra as a proportion of total	.314	.458	.498	.535	.542

Data are from Table 8.1 and author's estimates.

TABLE 9.5. THE IMPACT OF PETROLEUM REVENUES ON BIAFRAN
GOVERNMENT REVENUES, 1969–73
(*Million £ Nigerian, except as otherwise indicated*)

Revenues	1969	1970	1971	1972	1973
		LOW PROJECTION			
Petroleum revenues[a]	7	31	53	71	90
Nonpetroleum revenues[b]	45	50	54	57	60
Total revenues	52	81	107	128	150
Petroleum revenues as per cent of total revenues	13	38	50	55	60
		HIGH PROJECTION			
Petroleum revenues[a]	7	37	72	100	130
Nonpetroleum revenues[b]	45	50	54	57	60
Total revenues	52	87	126	157	190
Petroleum revenues as per cent of total revenues	13	43	57	64	68

Data are from Tables 8.4 and 9.4, and author's estimates.

[a] Calculated by applying Biafran proportions of total production from Table 9.4 to total petroleum revenues from Table 8.4.

[b] Calculated by assuming Biafra's proportion of total nonpetroleum revenues to be .30 and applying this to nonpetroleum revenues from Table 8.4.

TABLE 9.6. BIAFRAN PETROLEUM INDUSTRY
BALANCE-OF-PAYMENTS IMPACT, 1969–73
(Million £ Nigerian)

Item	1969	1970	1971	1972	1973
	LOW PROJECTION				
Payments to government	7	31	53	71	90
Other local payments[a]	10	19	21	23	25
Proceeds from local sales[b]	—	3	4	4	4
Balance-of-payments impact	17	47	70	90	111
	HIGH PROJECTION				
Payments to government	7	37	72	100	130
Other local payments[a]	10	22	28	32	34
Proceeds from local sales[b]	—	3	4	4	4
Balance-of-payments impact	17	56	96	128	160

Data are from Tables 8.5, 9.4, and 9.5, and author's estimates.

[a] Calculated by applying Biafran proportions of total production from Table 9.4 to total other local payments from Table 8.5.

[b] Calculated by assuming Biafra's proportion of total proceeds from local sales to be .30 and applying this to total proceeds from local sales from Table 8.5.

company payments to the Biafran government implied by these levels of production increase accordingly. If one applies the assumptions of Table 9.5 to the peak year, 1966, to establish a basis of comparison, Biafra would have benefited in that year from about £54 million of non-oil revenues and £7 million of oil revenues, a total of £61 million. On the low projection, petroleum revenues alone are half again this level by 1973, and they are more than double the 1966 level under the high projection. Burgeoning oil revenues would make up half to two-thirds of total government revenues and cause Biafran public receipts at least to double and possibly to triple within five years.

The petroleum industry's contribution to Biafran foreign exchange availability is tabulated in Table 9.6 and compared with non-oil availabilities in Table 9.7. Again referring to 1966 as a base, Biafra in that year might have had about £42 million of non-oil foreign exchange at its disposal (assuming that one-fifth of total non-oil Nigerian export earnings plus net capital flows is Biafran). To this should be added about £29 million of foreign exchange available from the Biafran share of the petroleum balance-of-payments impact. Owing almost wholly to petroleum contributions,

TABLE 9.7. PETROLEUM INDUSTRY CONTRIBUTION TO BIAFRAN FOREIGN
EXCHANGE AVAILABILITY AND USE, 1969–73
(*Million £ Nigerian, except as otherwise indicated*)

Item	1969	1970	1971	1972	1973
LOW PROJECTION					
Foreign exchange availability:					
Petroleum balance-of-payments impact	17	47	70	90	111
Nonpetroleum[a]	46	50	53	55	57
Total	63	97	123	145	168
Foreign exchange use[b]	63	97	123	145	168
Petroleum balance-of-payments impact as per cent of foreign exchange availability or use	*27*	*48*	*57*	*62*	*66*
HIGH PROJECTION					
Foreign exchange availability:					
Petroleum balance-of-payments impact	17	56	96	128	160
Nonpetroleum[a]	46	50	53	55	57
Total	63	106	149	183	217
Foreign exchange use[b]	63	106	149	183	217
Petroleum balance-of-payments impact as per cent of foreign exchange availability or use	*27*	*53*	*64*	*70*	*74*

Data are from Tables 8.8 and 9.6 and author's estimates.

[a] Exports plus net capital flows. Calculated by assuming Biafra's proportion of total nonpetroleum (exports plus net capital flows) to be .20 and applying this to total nonpetroleum (exports plus net capital flows) from Table 8.7.

[b] Nonpetroleum imports of goods and services, net transfer paid abroad, net increase in foreign exchange reserves.

which could rise to make up two-thirds to three-fourths of the total, this hypothetical figure of foreign exchange availability in 1966 might more than double (low projection) or triple (high projection) by 1973.

This contribution of foreign exchange would have been crucial for Biafra. It will be recalled that Nigeria has faced a foreign exchange constraint for the past several years. There are strong indications that the former Eastern Region was a large net user of foreign exchange within the Nigerian Federation. During the first half of the 1960's, Eastern Nigerian non-oil exports amounted to about 16–20 per cent of the Nigerian non-oil totals. Eastern Region non-oil imports are harder to estimate, but educated guesses

TABLE 9.8. THE CONTRIBUTION OF PETROLEUM VALUE ADDED TO
BIAFRAN GROSS NATIONAL PRODUCT, 1969–73
(*Million £ Nigerian, except as otherwise indicated*)

Value added	1969	1970	1971	1972	1973
	LOW PROJECTION				
Petroleum value added[a]	9	36	59	78	99
Nonpetroleum value added[b]	398	414	431	448	466
Gross National Product[c]	407	450	490	526	565
Petroleum value added as per cent of GNP	*2*	*8*	*12*	*15*	*17*
	HIGH PROJECTION				
Petroleum value added[a]	9	42	80	110	142
Nonpetroleum value added[b]	398	414	431	448	466
Gross National Product[c]	407	456	511	558	608
Petroleum value added as per cent of GNP	*2*	*9*	*16*	*20*	*23*

Data are from Tables 8.11 and 9.4 and author's estimates. All figures pertain to accounting concepts for national product.

[a] Calculated by applying Biafran proportions of total production from Table 9.4 to petroleum value added from Table 8.11.

[b] Calculated by assuming Biafra's proportion of nonpetroleum value added to GNP to be .27 and applying this to total nonpetroleum value added to GNP from Table 8.11.

[c] At market prices.

place them in the range of 25–30 per cent of the country total. Since the implied deficit was only partially offset by non-oil net capital inflows, petroleum foreign exchange contributions would have filled an important potential gap in the Biafran economy. The scarcity of foreign exchange might have continued to constrain growth of Biafran output for at least as long as it is expected to in united Nigeria, i.e. through 1970.

Direct benefits of petroleum in an independent Biafra are measured by the contribution of petroleum value added to Biafran Gross National Product. Petroleum value added for Biafra is found by taking Biafran shares of the Nigerian totals, as explained in Table 9.8. For a reference point, Biafran total GNP in 1966, calculated with Table 9.8 methodology, was perhaps £394 million, including £16 million of petroleum value added. Using the low projection, petroleum value added rises to one-sixth of Biafran GNP, helping to increase the total in 1973 to almost 40 per cent above the 1969 total. Switching to the high projection, the equiva-

The Politics of Nigerian Oil

lent figures are nearly one-fourth and roughly 50 per cent, respectively. The implied compound annual growth rates during 1969–73 are 8 per cent and over 10 per cent for the two projections. These amounts and rates would be even larger if the annual Biafran shares were estimated on the basis of the results of the input-output model. Because GNP accounting is used, adjusting for realized prices has no impact. Allowing for factor opportunity costs would lower the net contribution of petroleum only marginally.

In brief, an independent Biafra would have had a petroleum-dominated economy, with oil being responsible for as much as two-thirds of government revenues, three-fourths of foreign exchange earnings, and one-fourth of GNP. The potential for growth that petroleum might have imparted to some twelve million Biafrans is enormous. In view of this, there can be no question about the existence of a solid petroleum-based rationale to support Biafran noneconomic motivations for secession.

Biafran gains from petroleum would have been losses for Nigeria (less Biafra) to the extent that positive impacts from oil, otherwise redistributed, would have remained in Biafra. To measure the possible magnitudes of these losses, it is useful to contrast the major oil impacts in united Nigeria with those in Nigeria (less Biafra). There is no real need to build up a whole new set of tables, for the adverse impacts on united Nigeria can be seen readily by looking only at the 1973 figures. In each instance the stated range below refers to low and high projection estimates, respectively, contained in Chapter Eight.

In united Nigeria, oil production in 1973 might be between 1.8 and 2.4 million barrels per day, whereas comparable figures in Nigeria (less Biafra) are 0.9 to 1.1 million. In contrast to the £175–240 million figure for payments to government in united Nigeria, oil revenues would amount to £85–110 million in Nigeria (less Biafra). In united Nigeria the oil sector foreign exchange contribution would be £211–288 million, but in Nigeria (less Biafra) it would be £100–128 million. Finally, the oil sector contribution to GNP would be £193–262 million in united

Nigeria, but would diminish to £94–120 in Nigeria (less Biafra). The prospect of having the loss of Biafra, and especially of Biafran oil, cut total government revenues by 45 per cent, reduce foreign exchange availability by nearly 40 per cent, and decrease GNP by 30 per cent surely provided a strong economic incentive for the leaders of Federal Nigeria to wish to maintain a united Nigeria.

CHAPTER TEN

Implications for Policy

By this point there has been a substantial recounting of institutional details pertaining to petroleum and the Nigerian economy, a wide array of economic theories and tools of analysis, and a spate of empirical results delineating the impacts of petroleum in Nigeria. It is now imperative to examine directly what the implications are for future policy decisions. Clearly, the information and analytical results supplied in the foregoing discussion contain major policy significance for decision-makers in the Nigerian government and in the international petroleum industry. Discussion in this concluding chapter is therefore concerned with an explicit treatment of the most important policy issues encountered in the historical and projective analyses of this book. Examination of policies followed by the government and by the industry in the past provides insights into the choice among future policy alternatives.

Historical Impact of Petroleum

Direct private foreign investment is initially attracted by the demonstration of demand pressures that guarantee a market for the new industry's output at economical prices. Foreign demand for crude petroleum has been and is expected to remain very strong and stable. Nigerian crude has not and should not in the foreseeable future face any major problems on this account, though this naturally depends on the evolution of factors affecting head office decisions regarding the expansion of alternative sources of supply. For the demand pressures from abroad to re-

sult in economic growth, Nigerian factor owners must counter with a positive supply response. The analytical framework employed in this study is in effect addressed to the whole question of the influences that different forms of supply response might have. The direct impacts of a new industry as well as any contributions of factors or linkages are limited in type and in degree by the surrounding institutional framework and by the technological properties of the industry's production process. Before one can measure the growth-inducing effects of petroleum production, it is necessary to examine each of these related aspects in turn.

Because the Nigerian petroleum industry is foreign-owned, international institutional aspects are brought into play. For Nigeria the most important result on this score is that the large, vertically-integrated international petroleum companies have established high crude transfer prices and thus created unusually large economic rents for host country governments to share.* Domestically, the principal institutional aspect of oil production in Nigeria concerns industry-government relations, especially the industry's financial arrangements with government. The methods of production employed by a foreign-owned mining industry place limitations on the scope for direct and indirect impacts. The skill-intensive petroleum production process, which continually draws on corporate expertise for the latest technological innovations, results in large payments to the Nigerian government, moderate local payments in wages and salaries, very small use of local materials, and large employment of imported goods and services.

The petroleum industry makes an important direct contribution to Nigerian national income. Petroleum has recently accounted for about 3 per cent of GDP and 2 per cent of GNP. These figures are reduced by over one-half if one allows for the social opportunity costs of factors employed.

Two overlapping but useful categories are employed to exam-

---

* The concept of economic rent should not be confused with the rents that exploring and producing companies pay to the Nigerian government. These rental payments are wholly unaffected by the structure of crude oil prices.

Implications for Policy

ine mechanisms through which foreign-financed growth in one sector might have indirect effects on other sectors of the domestic economy. The first concerns the new industry's contribution to other sectors of scarce factors of production, especially foreign exchange, investment resources, and trained manpower. Recently Nigerian growth has been constrained by a shortage of foreign exchange. Increasing earnings of foreign exchange from petroleum exports have eased the continued pressure on Nigeria's balance of payments and allowed additional growth of output. In addition, the oil industry's payments to the Nigerian government potentially translate into public savings, pending expenditure allocations by the government. Though limited to date, the future magnitude of these payments to government should provide an important source of domestic saving and thus investment resources. The other major factor contribution, skilled labor, has been much less important than the supplying of foreign exchange and investment resources. The petroleum industry has not been particularly notable in training Nigerian manpower. It does undertake important training functions, but these are limited to a small group of Nigerians, most of whom remain in the petroleum industry.

The second category of indirect effects involves linkage effects, benefits or costs derived from intersectoral relationships between the petroleum sector and other sectors of the economy. Forward linkages associated with the petroleum sector have provided more stimulus to Nigerian industrial efforts than backward linkages, though neither type of investment linkage has been of great importance. In light of the limited amounts of local wages and salaries and Nigerian-controlled profits associated with petroleum production, final demand linkages have had only moderate impact. Technological linkages deriving from the oil industry's operations have surely not been indispensable to Nigerian growth efforts, although the construction of roads, investment in social overhead capital, and the training of labor deserve greater attention than casual observers generally give them.

In the linkage typology used in this study, there is an additional

category, fiscal linkages, whose effects are wholly parallel to those of the immediate linkages associated with the petroleum industry's activities. These effects result from the government's use of funds paid to it by the industry. The importance of this type of linkage is therefore initially a function of governmental policy decisions. Petroleum is expected to become the source of the majority of Nigerian government revenues in the near future. Naturally the larger the amounts of industry payments to government, the greater is the scope for fiscal effects. To date the most significant economic benefits of petroleum in Nigeria have resulted from the contributions of foreign exchange (Nigeria's scarce factor) and investment resources and from fiscal linkages.

In considering the economic benefits attached to the production of petroleum, it should be recalled that associated political implications might have carried high costs. Petroleum provided a strong economically based motivation in favor of Biafran secession, or conversely in favor of Nigerian unity, and thus must be implicated as a contributary factor in the Nigerian civil war. If the petroleum industry is to participate in any significant way in Nigerian growth, the divisive impacts of the civil war must be overcome. Stolper has pinpointed a crucial issue facing contemporary Nigeria: "Undoubtedly there exists a national Nigerian political consciousness. But tribal feelings and loyalties and antagonisms are strong. The desire to create a nation that is viable and independent of the outside world in general (and of the former metropolitan power in particular) is one thing. Overcoming inner distrust is another."[1] Unless a working relationship among all ethnic groups is established in the aftermath of the civil war, the future of petroleum in Nigeria could be relatively unimportant.

Future Impact of Petroleum

This section summarizes the quantitative estimates contained in Chapter Eight, thereby setting the stage for the policy discussion that follows. Negative external effects of the operations of the petroleum industry, such as environmental pollution, are important for the localities affected but have minor significance for

Implications for Policy

the economy. Unlike certain other examples of private foreign investment, there is no evidence that the activities of the oil companies in Nigeria have caused unemployment of previously employed local factors. And given the size and diversity of the future Nigerian economy, any factor price distortions introduced by petroleum are not expected to result in the serious disadvantages associated with strongly dualistic economies. The petroleum industry *per se* should not involve important economic costs, except in the unlikely event that it uses more skilled labor than it trains.

The benefits side of the petroleum balance sheet is quite a different story. Benefits associated with the production of petroleum involve both the direct contribution to Nigerian value added and the indirect contributions of factors and linkage effects. To sharpen the focus on policy, it is convenient to distinguish between effects that result from activities of the petroleum industry irrespective of government action, although government economic policies can alter their direction, and effects that result from explicit government decisions.

A tabulation of future economic gains of the first type begins with the direct contribution of the oil sector to Nigerian income, a not insignificant item under any accounting convention. Of the three factor contributions considered in this study, only the oil industry's furnishing of skilled labor does not depend directly on government policy. The magnitude of this effect is expected to be very small. Immediate linkages—investment, final demand, and technological—occur without action by the government, though government policy can affect their intensity. All three are almost certain to be of limited importance. In short, the petroleum industry's net direct contribution to national income will be the only significant economic effect occurring in the absence of specific action by the government.

The petroleum industry's indirect impacts on Nigerian growth depend primarily on its contributions of foreign exchange and investment resources and on fiscal linkages. These three types of effects coincide precisely with the second type of economic bene-

fits mentioned above. Stress on government policy in this study is thus indispensable. The petroleum industry's payments to the Nigerian government are an integral part of all three of these effects. The degree to which these payments lead to domestic saving and thus to investment resources depends on the government's expenditure policy. Payments to government are made abroad in convertible currency and thus have a direct influence on the oil sector's contribution of foreign exchange to the rest of the Nigerian economy. The use of foreign exchange is directly controlled by the government through its import policy.

Fiscal linkages are clearly tied to the industry's payments to government. Since fiscal linkages arise as a result of government expenditure of oil-related revenues and encompass all three types of linkages, the government has the power to allocate its oil-associated expenditures so as to create effects that will tend to extend or offset the effects associated with the oil industry's immediate linkages. Payments to government by the petroleum industry also increase budgetary flexibility and hence allow the possibility of additional benefits for the economy. The Nigerian public sector is thus likely to grow faster than the private sector. This underscores the fact that Nigerian government decision-makers are going to have to shoulder an ever increasing burden in making economic policy and in allocating government resources.

In brief, the outlook for the impact of petroleum in Nigeria is mixed. Likely economic costs are small and potential gains are large. Whether this potential will in fact be realized largely depends on government policies regarding the allocation of revenues received from the petroleum companies and the use of oil-contributed foreign exchange. This is especially so since government spending imparts fiscal linkages that could entail important economic gains. The production of oil could result in rapid economic growth of the Nigerian economy. But successful participation of the Nigerian government is a necessary ingredient. It is therefore quite clear that policies of the government will play a large role in determining the extent of future economic gains from petroleum.

Major Policy Issues

Private direct foreign investment in a less developed country is capable of providing the basis for significant domestic economic benefits. The magnitude of these gains is tempered by international and local institutions and by economic and technological considerations surrounding the foreign investor's activity. But only the government of the host country can properly compare the prospective economic gains from foreign investment with any associated political costs such as reduced national autonomy. An economist's task involves detailing all benefits and costs and suggesting possible economic and political implications of alternative policies.

Certainly a central and perhaps the most crucial policy issue facing the Nigerian government is the question of who should exploit the substantial reserves of crude petroleum in Nigeria. This question would not seem to be particularly pressing, since all major concessions in Nigeria have at least twenty years to run before expiration. But it is well known that nations often take extra-legal actions in situations of this kind if the benefits seem great enough to outweigh any international opprobrium that is likely to follow. The complex problem of control of an important economic resource is often intertwined with concern over possible loss of national sovereignty.

The question of whether or not a less developed country that is an important source of crude petroleum should acquire part or complete ownership of the production of this resource can be answered only after careful consideration of economic and extra-economic influences. One important political aspect concerns the nation's attitudes toward foreign investment and the degree to which the autonomy of the host government has been compromised as a result of the foreign investor's activities. The economic aspect centers on two comparisons: the relative profitability of the local government's investing in petroleum as opposed to its alternative investment opportunities, and the relative degree of taxation of the petroleum industry in the country in question as

compared with that of other major petroleum-producing nations whose supplies are competitive.

After allowances are made for production and transportation cost differentials, quality differences, and so forth, if the degree of taxation domestically is less than that abroad, then presumably the host government will desire to increase its rate of taxation until the differential disappears and to maintain equality as other producers increase taxes. If the rate of taxation is increased so as to become greater than that existing elsewhere, then the host government runs the risk of having the international petroleum companies shift their sources of supply from the country with the relatively high tax rate to other producing countries. Given the existing pattern of ownership, the host government is thus constrained from increasing its rate of taxation on what may be a very profitable industry.

Consider next the relative profitability of the local government's investing in petroleum production compared with its alternative investment opportunities. If investment in petroleum is clearly more profitable than investment elsewhere, and, moreover, if the government is constrained from increasing its tax rate on petroleum for the reasons mentioned, then the government will most likely want to invest in petroleum.* Depending on its refining, transporting, and marketing prospects and its ability to manage production, the local government may or may not desire to obtain part or complete ownership of the petroleum facilities in its country.

In a study of petroleum in Nigeria, both the economic and the political issues raised in the foregoing discussion are extremely difficult to analyze, since empirical knowledge is limited. If the petroleum industry can successfully apply pressures in its dealings with the Nigerian government, this condition limits Nigerian autonomy and entails some unmeasurable political costs.[2] Any use of extra-legal procedures by the petroleum companies in their

* This assumes that at least one major producing country has chosen not to tax petroleum as heavily as possible and that the price charged to the local government for purchase of equity in petroleum production is not set in a way that would capitalize all of the differential profits.

Implications for Policy

dealings with civilian or military governments in Nigeria has not been documented. A more important concern is the possible existence of wholly legal pressures emanating from the government's dependence on oil-related revenues. In Nigeria the petroleum industry's payments to government have not to date been important enough to afford the companies an opportunity for great political leverage. But the government's freedom from dependence on oil revenues is expected to diminish rapidly in the near future. In addition, Nigerian autonomy might have been reduced by foreign government pressures to promote the petroleum companies' desires indirectly. It is difficult to guess at the degree to which British, Dutch, American, French, or Italian policies toward Nigeria have been influenced by petroleum.[3] To the extent that the Nigerians have successfully withstood any pressures generated directly by the petroleum companies or indirectly by their home governments, Nigerian autonomy may in fact have been strengthened. To some degree this can be said to have occurred vis-à-vis certain of the companies during the struggles over the change to tax terms approximating those of the Organization of Petroleum Exporting Countries and the imposition of the Companies Decree.

It seems certain that the industry desires to retain its current amicable relationship with the Nigerian government, and hence to extend the present pattern of large Western-owned international corporations privately exploiting the Nigerian petroleum reserves. But to determine whether this is the best method of achieving economic gains for Nigeria, one must examine the two comparisons suggested in the introductory discussion of this section. It was established earlier (p. 24) that Nigeria has not yet negotiated tax terms fully equivalent to those operative in OPEC countries, though the difference now is much less than before the changes of 1967. Until the degree of taxation in Nigeria is changed to approximate more closely conditions in the other major petroleum-producing countries, purchase of part or all of any petroleum producing companies would probably not be economic.

Very little can be said about specific alternative investment opportunities in Nigeria compared with the possibility of invest-

ing in petroleum production. There are surely some projects with a higher rate of return than might be attained in petroleum, but until more is known along these lines it is not possible to analyze economically whether the Nigerian government should seek to obtain part or total ownership of petroleum production companies.

The government displayed a flexible position regarding part ownership of petroleum-producing companies in Nigeria in the Petroleum Decree 1969, issued in November 1969. In granting all new leases and licenses, the government has reserved for itself the option of part ownership at the discretion of the Commissioner of Mines and Power. The only existing arrangement between the government and an exploring and producing company is the agreement with Agip, whereby the government at a time of its own choosing can purchase up to 30 per cent of that company by assuming a 30 per cent share of past and future costs. The stipulation in the new decree may be a harbinger of future arrangements of this nature.

Most petroleum-producing countries desire large domestic refining capacities. The refinery question is further complicated by the preference for industry that is exhibited by decision-makers in many less developed countries. The possible establishment of additional refining capacity has long been a popular topic in Nigeria. Recently several international petroleum firms have approached the Nigerian government to discuss the construction of at least one new refinery (see pp. 93–94). But it is at this time unclear whether any of these firms intends to create a large enough refining capacity to turn Nigeria into a major exporter of petroleum products.

A second set of future policy issues, integrally related to the first, concerns the magnitude of the benefits from petroleum, assuming the Nigerian government and the industry as currently defined decide to continue to operate within the existing framework. The quantity of economic gains is initially a function of the levels of petroleum output and of the Nigerian tax arrangements concerning petroleum. Because tax rates have an important effect on

Implications for Policy

production levels, the Nigerian government's tax policy greatly affects the oil industry. As an intermediate step in the pursuit of other economic goals, the government presumably desires to maximize its income from petroleum.

The government has increased effective tax rates on petroleum production (see pp. 24–28). The change to OPEC terms in early 1967 shifted the tax arrangements in Nigeria to make them nearly equal to those existing in other major oil-producing underdeveloped countries. By changing its tax terms, the Nigerian government surrendered part of an anticipated increase in production rates to gain a large percentage increase in tax revenue per barrel. This increase is effectively about 25 per cent in years after each company's carried-over capital allowances are exhausted, assuming a transfer price of $1.85 per barrel under prior terms. Including only the years 1966–73 and employing the high projection of Chapter Eight, the additional payments resulting from the higher tax rate total nearly £130 million, with a 1966 present value of about £75 million (discounted at 10 per cent).*

Clearly any statements such as these depend on the realization prices that would have existed in the absence of a posted price. If the average transfer price for the industry were $1.95 per barrel, then an output one-sixth higher than the one currently projected would have been required for the government's income from petroleum to remain unchanged. Output would not have had to increase at all in this instance if OPEC terms had involved only the imposition of a posted price, since the effective posted price is $1.95 per barrel. But the OPEC terms have also required that royalties be treated as expenses rather than as tax offsets. A transfer price of $1.85 per barrel would require output over one-fourth greater than that now planned. At a transfer price of $1.75, the required increase in production would be nearly two-fifths, whereas at $1.65, it would be more than one-half.

This calculation has important policy implications. Many ana-

---

* This calculation assumes that carried-over capital allowances are available for all companies through 1970 only. See Appendix B for the equations employed.

lysts, especially those in the petroleum industry, argue that the Nigerian government's move to OPEC terms was badly timed, since some companies would have planned and invested for larger future production levels in Nigeria if the lighter tax terms had been retained longer. Analysis here has indicated that the required additional output would have to be as much as one-fourth to one-half above levels already planned. It is debatable whether or not such increases in output would have been forthcoming if the old tax terms had been retained. But evidence points to the conclusion that the Nigerian government's move to OPEC terms was certainly correct if slightly premature.

Two additional policy issues concerning the Nigerian government and the petroleum industry are given prominent mention in the Petroleum Decree 1969. Like the new concession arrangements, the new policies on the recruitment and training of Nigerians and the use of natural gas do not apply to operations in areas where companies are now exploring for and producing oil. But these policies might also be good indicators of the directions in which the Nigerian government desires to alter its relationship with the petroleum industry. The new decree states that all future recipients of oil mining leases must guarantee that within ten years they will employ Nigerian citizens in 75 per cent of management, professional, and supervisory positions and in 100 per cent of all other jobs. Moreover, holders of future oil prospecting licenses must submit a detailed program for recruitment and training of Nigerians within the first year of operation.

The new provisions concerning natural gas are less precise, but the government's concern is amply demonstrated. The Commissioner of Mines and Power reserves the right to include special provisions for natural gas in any future license or lease, and every new concessionaire must agree to submit feasibility studies for the utilization of natural gas within five years after beginning production. The new decree contains no reference to the price of natural gas, even though this is sure to become increasingly important as forward-linked industries become established.

On both of these issues, positive action by the petroleum indus-

Implications for Policy

try could enhance prospects for Nigerian economic development. It is clear that greater utilization of natural gas at as low a price as possible would help bring about the rapid creation of industries using gas as an input or as a source of energy. Moreover, the industry could train many more entrepreneurs, managers, technicians, and skilled laborers than it could use itself. Some of these newly skilled Nigerians could be employed in backward-linked industries that service or supply the petroleum exploring and producing industry and in forward-linked industries such as those that produce petrochemicals and manufacture fertilizers. Industry policy to supply skilled labor to other sectors of the Nigerian economy would increase the availability of the factor that should soon be scarce in Nigeria.

A final group of policy issues arises once the government has decided upon its tax arrangements with the petroleum industry and hence exercised its most important lever in the determination of economic benefits associated with the production of petroleum. This group of questions concerns the distribution of the gains from oil as well as the attainment of further gains. This subject naturally borders on the general subject of planning future allocations of investment and current expenditure of the government.[4]

In general, a large country like Nigeria has a wide spectrum of policy choices open to it in deciding how to guide private sector choices and how to use expanding revenues. As an economic ideal, the Nigerian government's economic policy decisions should reduce potential sectoral competition for economic resources and increase sectoral complementarity while its public expenditures should result in prudent social overhead and directly productive investments. But national income is not necessarily the proper welfare criterion unless its maximization is subject to a host of political as well as economic constraints.

Government policies should create as few economic disincentives as revenue needs allow. For example, it would seem to make sense to utilize some of the budgetary flexibility afforded by quickly increasing oil-derived government revenues to reduce the tax burden on the agricultural sector in general and on the agri-

cultural export sector in particular. By reducing or eliminating agricultural export taxes and decreasing marketing board profits, the government could turn the sectoral terms of trade in favor of the agricultural sector and thus increase agricultural output and possibly slow down the rate of rural-to-urban migration.[5] In addition, expenditures for education might be increased or reallocated in order to favor those types of activities that supply skilled labor to the economy (e.g. educating managers, training technicians). This need not necessarily occur at the expense of mass primary education programs, since petroleum revenues could allow simultaneous expansion of both primary and technical secondary education. Education policies will have an important role in determining future supplies of the factor of production that has been identified as the likely future constraint on faster economic growth.

In summary, private foreign investment in Nigerian petroleum has been responsible for net direct and indirect benefits that in aggregate amounted to about 7 per cent of prewar national income in Nigeria, a figure that might increase to perhaps 18 per cent of a much larger total during the next five years. Imputed political costs would have to be extraordinarily high to offset potential gains of this magnitude and cause a Nigerian government in the future to desire a major structural shift away from current arrangements for exploiting petroleum in Nigeria.

Appendixes

APPENDIX A

National Income and Balance of Payments

Six pages of tables follow.

TABLE A.1. NIGERIAN NATIONAL INCOME ACCOUNTS, 1950–68
(1950–64 IN CURRENT PRICES; 1965–68 IN 1964 PRICES)
(*Million £ Nigerian*)

Item	1950	1951	1952	1953	1954	1955	1956
1. Income:							
GDP at factor cost	512.1	573.2	614.5	665.0	774.2	827.5	870.6
Taxes less subsidies	12.2	13.8	16.3	17.9	20.6	23.7	29.4
GDP at market prices	524.3	587.0	630.8	682.9	794.8	851.2	900.0
2. Foreign trade:							
Exports:							
Goods	85.9	110.6	121.2	123.7	148.8	121.2	127.4
Services	3.0	3.6	5.1	4.9	5.5	7.0	8.3
Total exports	88.9	115.3	136.4	138.6	154.3	128.2	135.7
Imports:							
Goods	61.8	97.4	103.2	101.9	114.9	139.7	155.2
Services	4.2	6.1	7.2	9.9	11.0	12.6	13.9
Total imports	66.0	103.5	110.4	111.8	125.9	152.3	169.1
Net exports	22.9	10.7	15.9	16.8	28.4	−24.1	−33.4
3. Expenditures on GDP:							
Gross fixed capital formation:							
Public	...	11.3	17.2	19.2	25.9	28.5	34.0
Private	...	30.5	38.1	44.4	59.8	80.3	94.3
Total fixed capital	...	41.8	55.3	63.6	85.7	108.8	128.3
Consumption (residual):							
Public	...	24.1	229.6	34.8	38.1	58.3	67.3
Private (residual)	...	514.2	529.0	568.2	641.5	705.5	734.8
Total consumption	...	538.3	558.6	603.0	679.6	763.8	802.1
Total expenditures	524.3	587.0	630.8	682.9	794.8	851.2	900.0
4. Gross National Product:							
Factor payments made	5.9	7.0	7.8	8.3	7.3	7.3	6.0
Factor payments received	2.8	2.5	4.8	5.1	5.9	6.7	6.9
Net factor payments	3.1	4.5	3.0	3.2	1.4	0.6	− 0.9
GNP at market prices	521.2	582.5	627.8	679.7	793.4	850.6	900.9
5. Savings:							
Gross domestic saving	...	48.7	72.2	79.9	115.2	87.4	97.9
Gross national saving	...	44.2	69.2	76.7	113.8	86.8	98.8
Current account balance	21.3	6.9	16.9	16.3	29.5	−21.4	−30.4
6. GDP by subsector (factor cost):							
Agriculture & livestock	327.1	371.6	370.2	413.2	481.2	514.2	515.1
Domestic consumption	...	...	...	...	...	...	...
Exports	...	...	...	...	...	...	...
Fishing & forestry	13.8	17.5	15.2	16.6	18.8	22.0	26.6
Petroleum	—	—	—	—	—	—	—
Other	171.2	184.1	229.1	239.2	274.2	291.3	328.9
Mfg. & pub. utilities	3.3	3.5	4.9	6.4	7.1	12.2	8.4
Total GDP by subsector	512.1	573.2	614.5	665.0	774.2	827.5	870.6

TABLE A.1 (continued)

Item	1957	1958	1959	1960	1961	1962
1. Income:						
GDP at factor cost	910.0	864.9	903.0	918.9	1,049.2	1,147.6
Taxes less subsidies	28.7	45.3	52.5	64.0	67.6	73.0
GDP at market prices	938.7	910.2	955.5	982.9	1,116.8	1,220.6
2. Foreign trade:						
Exports:						
Goods	119.1	134.3	162.1	165.0	170.9	165.4
Services	10.0	10.1	10.7	11.6	16.7	18.0
Total exports	129.1	144.4	172.8	176.6	187.6	183.4
Imports:						
Goods	158.0	168.2	180.2	209.3	214.6	199.0
Services	17.6	18.6	22.9	30.4	27.0	29.1
Total imports	175.6	186.8	203.1	239.7	241.6	228.1
Net exports	−46.5	−42.4	−30.3	−63.1	−54.0	− 4.7
3. Expenditures on GDP:						
Gross fixed capital formation:						
Public	38.2	49.8	61.9	61.5	60.3	64.5
Private	61.1	59.4	60.9	67.6	92.2	95.3
Total fixed capital	99.3	109.2	122.8	129.1	152.5	159.8
Consumption (residual):						
Public	72.1	71.1	89.2	101.3	109.3	111.9
Private (residual)	808.3	770.9	778.5	821.8	916.7	998.2
Total consumption	880.4	842.0	867.7	923.1	1,026.0	1,110.1
Total expenditures	938.7	910.2	955.5	982.9	1,116.8	1,220.6
4. Gross National Product:						
Factor payments made	4.2	6.8	10.0	10.0	9.9	9.5
Factor payments received	8.4	7.5	7.0	5.5	6.0	5.6
Net factor payments	− 4.2	− 0.7	3.0	4.5	3.9	3.9
GNP at market prices	942.9	910.9	952.5	978.4	1,112.9	1,216.7
5. Savings:						
Gross domestic saving	58.3	68.2	87.8	59.8	90.8	110.5
Gross national saving	62.5	68.9	84.8	55.3	86.9	106.6
Current account balance	−41.0	−41.0	−35.0	−69.3	−61.7	−49.3
6. GDP by subsector (factor cost):						
Agriculture & livestock	537.3	567.3	547.6	529.9	621.2	677.4
Domestic consumption	...	...	...	...	...	...
Exports	...	...	...	...	...	...
Fishing & forestry	27.9	27.4	31.4	33.6	36.5	37.1
Petroleum	—	—	—	—	—	—
Other	344.8	270.3	324.0	355.4	391.5	433.1
Mfg. & pub. utilities	13.1	26.7	34.0	40.3	44.1	46.7
Total GDP by subsector	910.0	864.9	903.0	918.9	1,049.2	1,147.6

TABLE A.1 (*continued*)

Item	1963	1964	1965	1966	1967	1968
1. Income:						
GDP at factor cost	1,208.4	1,220.1	1,299.4	1,344.9	1,365.1	1,361.0
Taxes less subsidies	102.2	104.6	114.1	118.1	119.8	119.5
GDP at market prices	1,310.6	1,324.7	1,413.5	1,463.0	1,484.9	1,480.5
2. Foreign trade:						
Exports:						
Goods	186.0	211.0	265.3	279.1	239.2	208.4
Services	19.9	21.4	23.7	21.4	20.0	20.6
Total exports	205.9	232.4	289.0	300.5	259.2	229.0
Imports:						
Goods	200.6	246.2	268.1	248.0	214.6	189.2
Services	57.1	62.1	73.3	94.3	113.8	128.9
Total imports	257.7	308.3	341.4	342.3	328.4	318.1
Net exports	−51.8	−75.9	−52.4	−41.8	−69.2	−89.1
3. Expenditures on GDP:						
Gross fixed capital formation:						
Public	63.4	68.0	83.8	90.9	...	...
Private	113.6	127.0	150.3	151.7	...	...
Total fixed capital	177.0	195.0	234.1	242.6	237.2	231.1
Consumption (residual):						
Public	136.8	...	...	...	...	...
Private (residual)	1,051.3	...	...	...	...	...
Total consumption	1,188.1	1,205.8	1,240.4	1,257.7	1,312.7	1,332.7
Total expenditures	1,310.6	1,324.7	1,413.5	1,463.0	1,484.9	1,480.5
4. Gross National Product:						
Factor payments made	21.0	28.3	32.3	...	...	...
Factor payments received	4.0	3.5	3.0	...	...	...
Net factor payments	17.0	24.8	29.3	37.7	40.2	20.0
GNP at market prices	1,293.6	1,299.9	1,384.2	1,425.3	1,444.7	1,460.5
5. Savings:						
Gross domestic saving	122.5	118.9	173.1	205.3	172.2	147.8
Gross national saving	105.5	94.1	143.8	167.6	132.0	127.8
Current account balance	−54.5	−76.1	−49.7	−40.1	−65.0	−83.3
6. GDP by subsector (factor cost):						
Agriculture & livestock	677.0	626.5	663.8	669.1	673.4	676.7
Domestic consumption	577.6	521.0	547.0	561.0	575.0	583.0
Exports	99.4	105.5	116.8	108.1	98.4	93.7
Fishing & forestry	39.2	52.1	54.7	57.4	60.2	63.1
Petroleum	7.9	16.5	32.4	43.5	51.0	18.0
Other	484.3	525.0	548.5	574.9	580.5	603.2
Mfg. & pub. utilities	65.4	62.8	...	...	...	...
Total GDP by subsector	1,208.4	1,220.1	1,299.4	1,344.9	1,365.1	1,361.0

SOURCES AND NOTES FOR TABLE A.1

Sources:
Data are from Aboyade; Clark; Nigeria [15]; Helleiner, *Peasant Agriculture*; *Balance of Payments Yearbook*, XVII, XVIII (Washington, D.C.: International Monetary Fund, 1966, 1967); Lewis; P. N. C. Okigbo, *Nigerian National Accounts, 1950–57* (Lagos: Federal Ministry of Economic Development, 1962); H. M. A. Onitiri, "External Trade and Payments and Capital Flows" (Ibadan: NISER, 1969, mimeographed).

Notes:
General. The author is indebted to John B. Cownie, who collaborated with him in making this table; as the division of labor worked out, Cownie in fact did the lion's share of the work. "Years" are, in principle, Nigerian fiscal years, which run from April 1 of the calendar year named to March 31 of the next calendar year. However, balance-of-payments statistics (for calendar years) have often been used with no attempt at adjustment. Vertical bars on the table identify changes in the sources (or methods) used in generating the time series.

1. *Income.* Gross Domestic Product at factor cost: 1950–57, Okigbo; 1958–64, Nigeria [15]; 1965–67, estimated growth rates are 6.7 per cent (1964–65), 3.5 per cent (1965–66), 1.5 per cent (1966–67), and —0.3 per cent (1967–68).

Indirect taxes less subsidies: 1950–62, Clark; 1963–64, USAID estimate; 1965–68, assumed to grow proportionately with GDP at factor cost.

Gross Domestic Product at market prices: 1950–68 (GDP at factor cost) plus (indirect taxes less subsidies).

2. *Foreign trade.* Exports and imports of goods and services: 1950–62, Clark's estimates, based upon Helleiner and IMF; 1963–68, goods, derived from Table A.2; exports, services estimated as 12 per cent of non-oil exports, imports, services estimated as a residual using data from Table A.2.

3. *Expenditures on Gross Domestic Product.* Gross fixed capital formation, public and private: 1951–56, Aboyade; 1957–64, Nigeria [15]; 1965–68, calculated as the sum of estimated investment in the Petroleum sector plus all other investment.

Consumption, 1951–68: residual (GDP at market prices minus gross domestic saving), which therefore includes such items as inventory investment and statistical discrepancies. Consumption, 1951–63, public, Clark's estimates; private, residual.

4. *Gross National Product.* Net factor payments abroad, payments made and received: 1950–65; Clark's estimates, based upon Helleiner and IMF; 1966–68, Onitiri, p. 49.

Gross National Product at market prices, 1950–68, GDP at market prices minus net factor payments abroad.

5. *Savings.* Gross domestic saving, 1951–68, gross fixed capital formation plus current account balance.

Gross national saving, 1951–68, gross domestic saving minus net factor payments made abroad.
Current account balance: 1951–62, Clark's estimates, based upon Helleiner and IMF; 1963–68, Table A.2.

6. *Gross Domestic Product at factor cost by subsector.* All subsectors except Petroleum: 1950–57, Okigbo; 1958–64, Nigeria [15].

Agriculture and livestock subsector: exports, value added in 1963 is an updated estimate calculated from a figure derived by Lewis; value added is then presumed to change at the same rate as the changes in the total value of the principal export crops from 1964 to 1968. Domestic consumption, 1963–64, residual; 1965 assumed to grow 5 per cent; 1966–68, assumed to grow 1/2 per cent faster than a projected rate of population growth, or 2.6 per cent growth rate for 1966–68. The years 1965–68 include domestic consumption plus exports.

Fishing and forestry subsector, 1965–68, a 5 per cent growth rate is assumed; this is a compromise between a lower estimate by Lewis and the very rapid recent growth rates.

Petroleum subsector, 1950–62, included in Other; 1963–68, Table 6.2.

Other subsector, 1965–68, manufacturing and public utilities are residual.

TABLE A.2. NIGERIAN BALANCE OF PAYMENTS, 1963–68
(Million £ Nigerian)

Item	1963			1964			1965		
	Non-petroleum	Petroleum	Total	Non-petroleum	Petroleum	Total	Non-petroleum	Petroleum	Total
Visible trade:									
Exports of goods[a]	165.9	20.1	186.0	179.0	32.0	211.0	197.2	68.1	265.3
Imports of goods[a]	(196.4)	(4.2)	(200.6)	(234.5)	(11.7)	(246.2)	(254.6)	(13.5)	(268.1)
Trade balance	(30.5)	15.9	(14.6)	(55.5)	20.3	(35.2)	(57.4)	54.6	(2.8)
Invisible transactions (net)[b]	(30.6)	(6.6)	(37.2)	(27.4)	(13.3)	(40.7)	(27.2)	(36.4)	(63.6)
Balance on goods and services	(61.1)	9.3	(51.8)	(82.9)	7.0	(75.9)	(84.6)	18.2	(66.4)
Transfer payments (net)[a]									
Private[a]	(2.7)	—	(2.7)	(0.2)	—	(0.2)	2.7	—	2.7
Official[a]	(5.2)	—	(5.2)	(6.4)	—	(6.4)	(6.7)	—	(6.7)
	2.5	—	2.5	6.2	—	6.2	9.4	—	9.4
Balance on current account	(63.8)	9.3	(54.5)	(83.1)	7.0	(76.1)	(81.9)	18.2	(63.7)
Capital transactions (net):									
Private[c]	25.5	5.0	30.5	27.7	18.1	45.8	25.3	17.5	42.8
Official[a]	(3.3)	—	(3.3)	21.9	—	21.9	29.0	—	29.0
Balance on current and capital account	(41.6)	14.3	(27.3)	(33.5)	25.1	(8.4)	(27.6)	35.7	8.1
Errors and omissions[d]	(17.7)	—	(17.7)	(8.3)	—	(8.3)	3.8	—	3.8
Net change in official and banking assets (increase, if in parentheses)[a]	45.0	—	45.0	16.7	—	16.7	(11.9)	—	(11.9)

TABLE A.2 (continued)

Item	1966			1967			1968		
	Non-petroleum	Petroleum	Total	Non-petroleum	Petroleum	Total	Non-petroleum	Petroleum	Total
Visible trade:									
Exports of goods[a]	187.1	92.0	279.1	166.8	72.4	239.2	171.8	36.6	208.4
Imports of goods[a]	(228.5)	(19.5)	(248.0)	(200.9)	(13.7)	(214.6)	(173.3)	(13.6)	(186.9)
Trade balance	(41.4)	72.5	31.1	(34.1)	58.7	24.6	(1.5)	23.0	21.5
Invisible transactions (net)[b]	(34.0)	(57.9)	(91.9)	(56.3)	(56.5)	(112.8)	(73.5)	(34.8)	(108.3)
Balance on goods and services	(75.4)	14.6	(60.8)	(90.4)	2.2	(88.2)	(75.0)	(11.8)	(86.8)
Transfer payments (net)[a]	1.7	—	1.7	4.2	—	4.2	5.8	—	5.8
Private[a]	(6.3)	—	(6.3)	(4.5)	—	(4.5)	(2.8)	—	(2.8)
Official[a]	8.0	—	8.0	8.7	—	8.7	8.6	—	8.6
Balance on current account	(73.7)	14.6	(59.1)	(86.2)	2.2	(88.2)	(69.2)	(11.8)	(81.0)
Capital transactions (net):									
Private[c]	19.0	28.8	47.8	5.0	43.5	48.5	25.4	40.7	66.1
Official[a]	13.5	—	13.5	15.6	—	15.6	14.6	—	14.6
Balance on current and capital account	(41.2)	43.4	2.2	(65.6)	45.7	(19.9)	(29.2)	28.9	(0.3)
Errors and omissions[d]	(11.0)	—	(11.0)	(13.3)	—	(13.3)	(1.8)	—	(1.8)
Net change in official and banking assets (increase, if in parentheses)[a]	8.8	—	8.8	33.2	—	33.2	2.1	—	2.1

Data are from Table 6.8; Nigeria [1], p. 52; Nigeria [2], p. 76; "Private Foreign Investment in 1965," *Central Bank of Nigeria Economic and Financial Review*, V (December 1967), 1–7; and unpublished Central Bank estimates. All petroleum sector data are taken directly from Table 6.8, i.e. from unpublished company data and not from Central Bank estimates. All figures in parentheses are negative.

[a] All nonpetroleum data for lines indicated are taken from Central Bank *Annual Reports* and unpublished estimates.

[b] Nonpetroleum figure for 1963 is found as a residual from Central Bank total and Table 6.6 petroleum data; all other years' nonpetroleum figures are taken directly from Central Bank *Annual Reports* and unpublished estimates.

[c] Nonpetroleum figures for 1963–65 are estimated from data reported in Central Bank private foreign investment surveys, i.e. net private foreign capital inflow less increase in total mining investment for each year; 1966 through 1968 figures are preliminary estimates based on fragmentary evidence.

[d] Found as a residual.

APPENDIX B

Payments of Companies to the Government

The purpose of this appendix is to summarize the relevant details surrounding the calculation of the various payments that the petroleum exploring and producing companies make to the Nigerian government. This discussion is not wholly self-contained but must be read in conjunction with the discussion of financial arrangements in Chapter Two. This summary integrates pertinent provisions contained in the company covenants with the government and especially in the following Nigerian laws: Mineral Oil Act of 1 June 1958; Oil Pipelines Act of 1 June 1958, amended by Oil Pipelines Act of November 1965; Petroleum Profits Tax Ordinance, 1959; Income Tax (Amendment) Decree No. 65, 1966; and Petroleum Profits Tax (Amendment) Decree No. 1, 1967. No attempt is made to attribute individual provisions to particular laws, except with the two most recent changes. Many of the earlier legal details summarized here are elaborated in an explanatory document, "Shell-BP's Financial Arrangements with Government" (Port Harcourt, Nigeria: The Shell-BP Petroleum Development Company of Nigeria Limited, 1964). The principal provisions of Petroleum Decree 1969, Decree No. 51 of 1969, are summarized in the final sections of this appendix.

Types of Concession Arrangements

There are three types of legal permission that a petroleum company can receive from the Nigerian government to undertake activity in Nigeria. Oil Exploration Licenses (OEL's), valid for periods of one to two years, confer nonexclusive rights to make geological and geophysical studies but not to drill for oil. Oil Prospecting Licenses (OPL's) involve an obligation on the part of the

company to meet certain minimum drilling requirements. Moreover, they grant exclusive right for exploration and retention of oil discovered. Onshore OPL's have generally been granted for four years, with a one-year renewal option. Continental Shelf OPL's (CSOPL's) have been granted for four years, with two successive three-year extension options, the first automatic and the second negotiable. Upon expiration, no more than half of an OPL can be converted into Oil Mining Leases (OML's), with the other half of the acreage reverting to the government. Onshore OML's are valid for thirty years, with an option to renew for a second thirty years, whereas offshore OML's are valid for forty years. Premiums are attached to the granting of any or all of these concession arrangements, the exact amount being set by the government according to what the market will bear at the time.

Rentals

Rentals are specific taxes based wholly on the type of concession arrangement and the amount of acreage involved. All rentals are paid annually in advance each December. Onshore OPL's have generally carried a rate of five shillings ($0.70) per square mile or one pound ($2.80) on renewal, and offshore OPL's have paid two shillings and one pound respectively. OML fees on the other hand are much higher (both onshore and offshore) and are based on the following sliding scale: 2/6 (two shillings and sixpence) per acre in the first year, 3/6 per acre in the second, 4/6 per acre in the third, 6/– per acre in the fourth, 8/– per acre in the fifth, and 10/– per acre in the sixth and all succeeding years of the concession life.

Royalties

Royalties are ad valorem taxes based on the value of each barrel of crude oil produced. Prior to the 1967 change, each company calculated royalties on the basis of realized prices. As will be recalled from Chapter Two, one of the main features of the imposition of terms of the Organization of Petroleum Exporting Countries (OPEC) was the posting of a previously agreed-upon price for the calculation of gross proceeds from exported crude. Royalties are now equal to a certain percentage of field storage value, with the latter built up by multiplying output by the posted price

and deducting pumping and storage costs. Onshore the royalty rate is 12.5 per cent of field storage value. Pumping and storage costs might amount to 0.2–0.3 per cent of proceeds for most companies in the early stages of production and then decrease. Thus the effective royalty rate onshore, 12.2 per cent of gross proceeds, might eventually rise to about 12.4 per cent. Offshore, after a minimum rate of one-half million tons of crude per annum is achieved, the royalty rate is 10 per cent of field storage value between 3 and 10 miles offshore (zero to 3 miles is considered onshore) and 8 per cent beyond 10 miles offshore. The royalty rates for natural gas (which apply only to gas sold) are two pence per thousand cubic feet of gas sold onshore, and 4 per cent or 3.2 per cent of the sale price in the inner or outer offshore areas respectively. Royalty payments due are reduced by the amount of rentals paid on acreage in which production has taken place; for this calculation rental offsets accrue from all OML's that are a part of the former OPL in which production has occurred. The timing of the royalties payments is as follows: in August royalties (less all rental offsets) are paid on estimated first half-year production; in February of the next year the final royalty payment is made.

Petroleum Profits Tax

Compared to the figuring of rentals and royalties, the calculation of profits tax is a relatively complicated procedure. As with royalties, gross proceeds for exported crude are found using output times the posted price. The effective posted price used in finding gross proceeds will be smaller than that used in calculating royalties to the extent of allowable discounts, including percentage discounts, gravity differentials, and marketing allowances (see the accompanying discussion in Chapter Two). From gross proceeds one subtracts general operating and intangible appraisal and development drilling costs. As a consequence of the change to OPEC terms, royalties are also subtracted, for they now must be expensed, before arriving at assessable profit. Capital allowances (depreciation) are then subtracted from assessable profit to find chargeable profit, but their use is restricted. (The generation of capital allowances is discussed below.) As long as it has sufficient capital allowances available, each company is allowed to

offset all but 15 per cent of potential profits tax. Under the 50/50 profit-sharing principle, half of chargeable profit is paid to the government and half is retained by the company. From the government's half, known as assessable tax, one must subtract all tax offsets to arrive at chargeable tax or the petroleum profits tax that must be paid each year. Tax offsets comprise rentals (not already absorbed as offsets against royalties), customs duties, stamp duties, and other minor taxes paid by the companies for which no service was performed in return. Before the OPEC terms decree, royalties were included in tax offsets as well. Then the total payments to government equaled assessable tax, i.e. chargeable tax plus tax offsets. Now, with the expensing of royalties, total payments to government equal assessable tax plus royalties. With the assistance of some algebra, summary equations are developed below to show the impact of the OPEC terms under different assumptions regarding the availability of capital allowances. The timing of profits tax payments is significant, for it involves about a six-month lag: the first three installments of 25 per cent each of the estimated chargeable tax for the production year are made in September and December of that calendar year and March of the following year; the final payment is due within 21 days after the final assessment, i.e. late June or July of the year following the production year.

Harbor Dues and Port Charges

Petroleum companies that export oil are required to pay harbor dues and port charges to the Nigerian Ports Authority (a quasi-government statutory corporation). The rates for shipments from Bonny are three shillings and four pence per ton of crude shipped plus one shilling and eight pence per net registered ton paid by each tanker entering an export terminal (the latter works out to about eight pence per ton of crude oil exported). On a per-barrel basis these rates translate to about $0.06 per barrel harbor dues and $0.016 per barrel port charges. In May 1969 the government introduced the Offshore Terminal Dues Decree, retroactive to January 1968, which enables the Nigerian Ports Authority to charge a fee for services provided at offshore terminals, but to date no regulations have been issued.

Generation of Capital Allowances

The generation of capital allowances and their use over time is a highly significant part of the Nigerian petroleum industry's financial arrangements with the government. These allowances are used as permitted under the 15 per cent rule to decrease the amount of chargeable profit, i.e. that amount of which 50 per cent is paid to the government. Prior to the changes embedded in Decree No. 65 of 1966, capital allowances were generated for the indicated categories under the following depreciation schedules: qualifying building expenditure, e.g. houses, offices, warehouses, jetties, roads—initial 20 per cent, annual 10 per cent; qualifying plant expenditure, e.g. machinery, flow station equipment, pipelines, storage tanks, vehicles—initial 40 per cent, annual 20 per cent; qualifying petroleum expenditure, e.g. exploration surveys, exploration drilling—initial 25 per cent, annual 15 per cent.

No capital allowances were generated until the first accounting period during which petroleum production occurred. All previous capital expenditure was assumed to have taken place during that accounting period. The initial allowances were calculated by category. Then the annual allowances were figured on the basis of the amounts written down within each category. For example, if £10 million of qualifying petroleum expenditure had been made prior to and during the first production accounting period, the company could claim initial allowances of £2.5 million (25 per cent of £10 million) and annual allowances of £1.125 million (15 per cent of £7.5 million, which itself is £10 million less the £2.5 million already depreciated). In the second accounting period, initial allowances were taken on new capital expenditure. The new investment was written down; what remained was added to the written-down investment carried over from the previous year. Continuing the example, if, in the second year, £2 million of qualifying petroleum expenditure had been made, then the initial allowance on this amount would have been £0.5 million. The annual allowance was then calculated on a written-down investment of £7.875 million, or £6.375 million carried over from the first accounting period plus £1.5 million from the current year. This would have amounted to 15 per cent of £7.875 million, or £1.181 million, leaving £6.694 million to be carried over to the third ac-

TABLE B.1. AN EXAMPLE OF THE GENERATION OF CAPITAL ALLOWANCES
(Million £ Nigerian)

				Written-down investment	
Year	New investment	Capital allowances	On new investment	Before annual allowance	After annual allowance
One					
Initial	10.000	2.500^a	7.500	7.500	
Annual		1.125^b			6.375
Total		3.625			
Two					
Initial	2.000	.500^a	1.500	7.875	
Annual		1.181^b			6.694
Total		1.681			
Cumulative	12.000	5.306			6.694

^{a} At 25 per cent of new investment.
^{b} At 15 per cent of written-down investment before annual allowance.

counting period. This example can be summarized in Table B.1. The total amount of capital allowances generated in the two years is £5.306 million, while the written-down investment is £6.694 million. The sum of these two is £12.000 million, or the amount of total investment undertaken. Since the annual allowances were claimed on the basis of the written-down value, the capital assets were never completely depreciated until they were scrapped. When this occurred, the remaining balances could be claimed in full.

Under the former legislation, if what might be considered a normal dollar's worth of capitalized expenditure were split among the three categories, the weighted overall depreciation rates might be about 30 per cent for initial allowances and 15 per cent for annual allowances. The changed rates included in the 1966 amendment are as follows: qualifying building expenditure, 10 per cent and 5 per cent; qualifying petroleum expenditure, 15 per cent and 10 per cent; and qualifying plant expenditure, 20 percent and 10 per cent (with the annual allowance under the latter negotiable upward from a floor of 10 per cent). If the original weights are now applied to these amended depreciation rates, a normal dollar's worth of capitalized expenditure would be depreciated at an initial rate of about 15 per cent and an annual rate of about 10 per cent.

Use of Capital Allowances

The Nigerian Petroleum Profits Tax Ordinance of 1959 stated that the amount of restricted capital allowances that can be taken during any one year is equal to 85 per cent of the assessable profit less 170 per cent of the tax offsets. To derive this formula, the following symbols are adopted:

GPd = Gross proceeds
$(GOC + ADC)$ = General operating costs plus intangible appraisal drilling costs
AP = Assessable profit
RKA = Restricted capital allowances
CP = Chargeable profit
AT = Assessable tax
TO = Tax offsets
CT = Chargeable tax

Reducing what is stated above to symbols gives the following:

$GPd - (GOC + ADC) = AP \qquad .50\,CP = AT$
$AP - RKA = CP \qquad AT - TO = CT$

At least 15 per cent of the difference between 50 per cent of the assessable profit and the tax offsets must be paid as chargeable tax (assuming no previous operating losses and no carried-over tax offsets). Stated symbolically: $X = .50\,AP - TO$. The law requires that $CT = .15\,X$. Thus, $CT = .15\,(.50\,AP - TO)$.

The 15 per cent rule can now be arrived at by means of straightforward algebra. Since $AT - TO = CT$ and $.50\,CP = AT$, it follows that $CT = .50\,CP - TO$. But $CP = AP - RKA$; hence $CT = .50\,(AP - RKA) - TO$. Since also $CT = .15\,(.50\,AP - TO)$, we have

$$.15\,(.50\,AP - TO) = .50\,(AP - RKA) - TO$$

Multiplying by two,

$$.30\,(.50\,AP - TO) = (AP - RKA) - 2\,TO$$
$$.15\,AP - .30\,TO = AP - RKA - 2\,TO$$
$$RKA = AP - .15\,AP - 2\,TO + .30\,TO$$
$$= .85\,AP - 1.70\,TO$$

Payments of Companies to the Government

The algebra is carried a step further, in order to find the relationship between restricted capital allowances and chargeable tax. Given $CP = 2\,AT$ and $CP = AP - RKA$, it follows that $2\,AT = AP - RKA$, or $AP = 2\,AT - RKA$. Substituting gives

$$\begin{aligned} RKA &= .85\,(2\,AT - RKA) - 1.70\,TO \\ &= 1.70\,AT - .85\,RKA - 1.70\,TO \\ .15\,RKA &= 1.70\,AT - 1.70\,TO \\ &= 1.70\,(AT - TO) \end{aligned}$$

But $CT = AT - TO$; hence

$$\begin{aligned} .15\,RKA &= 1.70\,CT \\ RKA &= \frac{1.70\,CT}{.15} = 11.33\,CT \end{aligned}$$

Thus $CT = RKA/11.33$.

In summary, this is a system of five equations and eight variables:

(1) $\qquad GPd - (GOC + ADC) = AP$
(2) $\qquad\qquad\qquad AP - RKA = CP$
(3) $\qquad\qquad\qquad\quad AT - TO = CT$
(4) $\qquad\qquad\qquad\qquad .50\,CP = AT$
(5) $\qquad\qquad\qquad\qquad RKA = 11.33\,CT$

Equations (1) through (3) are definitional, equation (4) expresses the 50/50 profit division, and equation (5) results from the 15 per cent rule. If three of the variables are known (including either GPd or $(GOC + ADC)$ but not both and not both RKA and CT), then the remaining five can be calculated. It must be stressed that the system assumes that all available capital allowances are used as early as is legally possible. Furthermore, the system is no longer applicable after a company exhausts its carried-over capital allowances, i.e. RKA obviously cannot exceed the amount of available capital allowances. When carried-over allowances are used up, then the company naturally uses only the amounts that its new and previous investments generate annually. In this case, equation (5) would be inoperative and the system could be solved only if four variables (including either GPd or $(GOC + ADC)$ but not both, and either TO or CT but not both) were known.

The Revenue Implications of OPEC Terms

In Chapter Two the OPEC terms that were incorporated into Nigerian financial arrangements in 1967 were summarized. Here four formulas are developed to allow quick measurement of the financial impact of this change on government revenues. Naturally the imposition of OPEC terms has provided an economic disincentive with respect to future production levels, but in the absence of knowledge allowing a better assumption, the comparisons here assume production levels to remain constant. The same algebraic notation as in the previous section is employed, with the embellishment that all symbols with the subscript 1 refer to the former legislation and all symbols with the subscript 2 refer to OPEC terms (e.g. GPd_1 = gross proceeds under the former legislation, whereas GPd_2 = gross proceeds evaluated at a posted price under the current legislation).

First, the effect of expensing royalties (as opposed to using royalties as tax offsets) is examined in the situation where accumulated capital allowances have been exhausted. Ro_1 as a part of TO_1, i.e. under former arrangements, has a tax offset value of only half of its value. It is true that expensing Ro_1 decreases AT_1 by half of the value of Ro_1, everything else held constant. But under the old terms, AT_1 was equal to TGT_1, whereas under the current expensing set-up, $TGT_2 = AT_2 + Ro_2$. Thus with the changed terms, half of Ro_2 is subtracted from AT_2, but all of Ro_2 is added to AT_2 in the end, leaving the net change equal to $+ 1/2\, Ro_2$.

Things get slightly more complex when a posted price is added to the expensing arrangement. The increment in payments to government is made up of two parts. First of all, one-half of Ro_2, the new royalties figure, is added on the reasoning employed above. In addition, the gross proceeds total is increased by the setting of a posted price. Since capital allowances and costs remain constant, exactly one-half of the difference between GPd_2 and GPd_1 must be paid in taxes. Putting these two effects together yields

$$TGT_2 - TGT_1 = \frac{Ro_2}{2} + \frac{(GPd_2 - GPd_1)}{2}$$

Payments of Companies to the Government

$$\text{Thus } TGT_2 = TGT_1 + \frac{Ro_2}{2} + \frac{(GPd_2 - GPd_1)}{2}$$

where TGT_2 is the total government take under the full OPEC terms and TGT_1 is the total government take under past terms. To sum up, in order to find the increase in TGT_1 with the OPEC terms after accumulated capital allowances have been exhausted, add half of the OPEC royalty figure plus half of the difference between gross proceeds under the two arrangements to prior total government take.

It is somewhat more difficult to find the change in payments to government when the company still has enough capital allowances available to be able to take its legal limit for the year. The first case with these assumptions is straightforward, i.e. comparing the payments to government as between pre-OPEC terms and expensing of royalties only, given sufficient capital allowances. An expensed Ro_2 is again worth only half as much in tax offset value as an Ro_1 that is part of TO_1. But in light of the fact that the legal limit of capital allowances is claimed, the tax increase is not one-half of Ro_2, but only 15 per cent of one-half of Ro_2, or $.075\,Ro_2$. A quick review of the 15 per cent rule will make the reason for this immediately apparent. Thus the formula for finding the revenue impact of expensing royalties under an assumption of sufficient capital allowances is merely to multiply the amount of net royalties paid by .075. In terms of the symbols, $TGT_2 = TGT_1 + .075\,Ro_1$.

Unfortunately the picture gets slightly more complicated in the fourth and final case, i.e. where financial arrangements call for the expensing of royalties and the establishment of a posted price at a time when the company in question has ample capital allowances on hand to offset the adverse effect of the change to the greatest extent legally possible. The formula is made up of three parts. First of all, there is the increment of Ro_2 less Ro_1. With a posted price, the amount of royalties will be greater. Second, there is the effect of expensing the royalties, or .15 of $Ro_2/2$. The third and final part of the formula concerns the difference between assessable profits with the new terms and assessable profits with the expensing only terms: $AP_2 = GPd_2 - Ro_2 - (GOC_2 + ADC_2)$, where AP_2 is assessable profits under full OPEC terms;

and $AP_1 = GPd_1 - Ro_1 - (GOC_1 + ADC_1)$ under the expensing only terms; thus $AP_2 - AP_1 = GPd_2 - Ro_2 - GPd_1 + Ro_1$. Fifteen per cent of half of the difference between AP_2 and AP_1, or

$$\frac{.15}{2}(GPd_2 - Ro_2 - GPd_1 + Ro_1)$$

is required. Putting the three parts of the formula together gives:

$$TGT_2 - TGT_1 = (Ro_2 - Ro_1) + .15\frac{Ro_1}{2} +$$

$$\frac{.15}{2}(GPd_2 - Ro_2 - GPd_1 + Ro_1)$$

$$= Ro_2 - Ro_1 + \frac{.30}{2}Ro_1 - \frac{.15}{2}Ro_2 +$$

$$\frac{.15}{2}(GPd_2 - GPd_1)$$

$$= .925\,Ro_2 - .85\,Ro_1 + .075\,(GPd_2 - GPd_1)$$

Hence $TGT_2 = TGT_1 + .925\,Ro_2 - .85\,Ro_1 +$
$$.075\,(GPd_2 - GPd_1)$$

The formula in its final form shows that TGT_2 can be calculated by summing the total government take under pre-OPEC arrangements plus .925 of the royalty payment due under full OPEC arrangements less .85 of the royalty due under pre-OPEC arrangements plus .075 times the difference between the gross proceeds calculated at the posted price and the gross proceeds calculated at the pre-OPEC price.

Summing up, four formulas have been derived. Because Nigeria has adopted posted prices as well as the expensing of royalties, formulas (1) and (3) are best viewed as transitions to formulas (2) and (4), respectively. In each case TGT_2 is the total government take under the OPEC arrangements.

No accumulated capital allowances, royalties expensed:

(1) $$TGT_2 = TGT_1 + \frac{Ro_1}{2}$$

No accumulated capital allowances, royalties expensed and posted price established:

(2) $$TGT_2 = TGT_1 + \frac{Ro_2}{2} + \frac{(GPd_2 - GPd_1)}{2}$$

Ample accumulated capital allowances, royalties expensed:

(3) $\qquad TGT_2 = TGT_1 + .075\, Ro_1$

Ample accumulated capital allowances, royalties expensed and posted price established:

(4) $\qquad TGT_2 = TGT_1 + .925\, Ro_2 - .85\, Ro_1 + .075\,(GPd_2 - GPd_1)$

At the present time formula (4) is the applicable one. In the near future, however, as producing companies reach higher rates of production and exhaust their accumulated capital allowances, formula (2) will be relevant.

Major Provisions of Petroleum Decree 1969

In November 1969 the federal military government issued a new petroleum decree the provisions of which appear to apply only to new concessions and concessionaires. To date there have been no regulations applying any of these changes to existing arrangements. The decree is of interest in that it will definitely pertain to all new concessions, including the offshore areas that were relinquished in November 1968, and because it offers a good indication of government attitudes regarding the future course of its arrangements with private petroleum companies in Nigeria.

Major changes or innovations incorporated in the new decree include the following. First, Oil Mining Leases will not exceed 20 years, and after 10 years half the acreage of the OML will have to be relinquished. This compares with the earlier provisions for OML's of 30 years onshore and 40 years offshore with no relinquishment; upon conversion from OPL's to OML's, at least half the acreage has to be relinquished but none of the OML's must be given up. Second, if it is deemed to be in the public interest, the granting of any license or lease may be made contingent upon participation by the federal military government on terms to be negotiated with the applicant. Third, special provisions may apply to any natural gas discovered. But in any event, within five years after beginning production, a concessionaire must submit a program for the utilization of natural gas discovered whether associated with petroleum or not. Fourth, the holder of an OML must guarantee that within 10 years Nigerian citizens

will hold at least 75 per cent of the total managerial, professional, and supervisory positions, including at least 60 per cent of each individual grade, and that all skilled, semiskilled, and unskilled workers are Nigerian citizens. Moreover, holders of OML's must submit detailed programs for training and recruitment of Nigerians upon application for the lease, whereas holders of OPL's must do so within one year. Finally, the rates charged for rents and royalties are increased over earlier provisions and the timing of payments of royalties is altered. The new rates for rents are as follows: OEL's, £250 per year; OPL's, £1 per square mile; and OML's, £0.5 per acre during the first ten years and £1 per acre thereafter. Royalty rates are $12\frac{1}{2}$ per cent of field storage value for crude petroleum and 10 per cent of the price received for natural gas, irrespective of whether the production takes place onshore or offshore. Royalties are to be paid not more than one month after the end of every quarter, rather than twice per year as under earlier provisions.

APPENDIX C

Subsidiary Firms of the Nigerian Petroleum Industry

In June and July 1966 Wilson E. Schmidt and the author, while serving as economic consultants to the U.S. Agency for International Development Mission to Nigeria, undertook a survey of the firms that acted during 1965 as suppliers or contractors for the Nigerian petroleum exploring and producing industry. The quantifiable information obtained in this survey is summarized in Tables C.1 of this appendix and 6.8 of Chapter Six (p. 82). Much of this data is used in portions of the text in Chapters Five, Six, and especially Seven.

The method employed in completing this survey is described here so that the reader may better appraise the degree of reliability that he might wish to attach to the survey data. This is in no way meant as an apology for using self-gathered data in the absence of official statistics. Analogous data gathering and processing procedures by official agencies generally suffer from shortcomings similar to those encountered in this private survey.

In brief, the following method was employed to obtain information about the petroleum industry's suppliers and contractors in Nigeria. First, data regarding the exploring and producing companies' payments in Nigerian currency to their suppliers and contractors were obtained directly from each of the companies individually. One caution should be observed at this point. All of the companies had easily justifiable difficulties in making a clean division between their suppliers as opposed to their contractors, since many ancillary firms undertook both functions. Nonetheless the companies were able to offer information regarding the types of services performed by the contractors and the types of materials furnished by the suppliers. There was little to choose

TABLE C.1. SUMMARY OF INFORMATION RELATING TO INTERVIEWED FIRMS THAT SUPPLY OR SERVICE THE NIGERIAN PETROLEUM EXPLORING AND PRODUCING INDUSTRY, 1965 DATA
(*Thousand £ Nigerian, except as otherwise indicated*)

Firms	(1) Receipts from oil industry paid locally	(2) Receipts from oil industry paid abroad	(3) Receipts from all business paid locally	(4) Oil receipts/total receipts ratios (1) ÷ (3)	(5) Amount of (1) expatriate-owned
Suppliers	2,735	—	14,874	.184	2,713
Contractors					
Well-site preparation and roads	2,781	—	10,728	.259	2,361
Dredging	470	—	970	.485	470
Drilling	2,647	5,956	2,647	1.000	2,647
Surveys	1,088	956	1,088	1.000	1,088
Construction	1,066	2,417	18,320	.058	902
Catering	209	—	285	.733	150
Transport	2,200	750	7,707	.286	1,970
Total contractors	10,461	10,079	41,745	.251	9,588
Total suppliers and contractors	13,196	10,079	56,619	.233	12,301

TABLE C.1 (*continued*)

Firms	(6) Amount of (1) large Nigerian-owned	(7) Amount of (1) small Nigerian-owned	(8) Local intermediate inputs into (1) + (2)	(9) Imported intermediate inputs into (1) + (2)	(10) Value added in (1) + (2)
Suppliers	15	7	154	1,506	1,075
Contractors					
Well-site preparation and roads	420	—	254	817	1,710
Dredging	—	—	12	108	350
Drilling	—	—	265	870	7,468
Surveys	—	—	136	241	1,667
Construction	140	24	531	648	2,304
Catering	59	—	68	51	90
Transport	205	25	218	582	2,150
Total contractors	824	49	1,484	3,317	15,739
Total suppliers and contractors	839	56	1,638	4,823	16,814

Subsidiary Firms of the Nigerian Petroleum Industry

TABLE C.1 (continued)

Firms	(11) Oil-related gross capital investment	(12) Oil-related Nigerian employment (number employed)	(13) Oil-related wages and salaries	(14) Oil-related expatriate employment (number employed)	(15) Oil-related company taxes	(16) Oil-related Nigerian PAYE taxes
Suppliers	908	503	160	44	420	3
Contractors						
Well-site preparation and roads	816	2,637	422	73	9	4
Dredging	680	343	85	36	—	2
Drilling	7,811	1,294	417	320	392	18
Surveys	580	3,459	225	136	43	21
Construction	2,252	1,458	259	183	94	4
Catering	51	1,294	128	13	—	1
Transport	1,503	1,313	261	70	31	11
Total contractors	13,693	11,798	1,797	831	569	61
Total suppliers and contractors	14,601	12,301	1,957	875	989	64

TABLE C.1 (continued)

Firms	(17) Oil-related expatriate PAYE taxes	(18) Oil-related capital/output ratios (11) ÷ [(1)+(2)]	(19) Oil-related capital/value added ratios (11) ÷ (10)	(20) Oil-related input/output ratios (8) ÷ (9)	(21) Actual producing and exploring industry total local payments[a] (1)+(2)	(22) Survey blowup factor (21) ÷ (1)
Suppliers	48	.332	.845	.607	3,903	1.427
Contractors						
Well-site preparation and roads	24	.293	.477	.385	3,086	1.110
Dredging	21	1.447	1.943	.255	522	1.111
Drilling	138	.908	1.046	.132	3,000	1.133
Surveys	39	.284	.348	.184	1,183	1.087
Construction	109	.647	.977	.339	1,395	1.309
Catering	—	.244	.567	.569	230	1.100
Transport	12	.509	.699	.271	2,474	1.125
Total contractors	343	.667	.870	.234	11,890	1.137
Total suppliers and contractors	391	.627	.868	.278	15,793	1.197

[a] Figures are aggregations of data obtained separately from the producing and exploring companies.

from regarding materials, so the local purchases of goods were left intact as one category. The breakdown of contractors into the seven categories shown in Table C.1 was fashioned on the advice of officials of the exploring and producing companies.

The next step was to interview the supplying and servicing firms whose names were received from the exploring and producing companies. Since the list was much too long for a complete census to be done in the time available, it was decided to interview as many as possible of the large firms, i.e. those with oil company receipts of more than £15,000 in 1965, and in addition to interview a selected sample of small firms, i.e. those with earnings in 1965 of less than £15,000. To maintain internal consistency and to carry out cross checks, Schmidt and the author did all of the interviews personally. Generally the ancillary firm representative interviewed was the general manager, often aided by his accountant. Of the nearly eighty firms covered in the survey, only two willfully withheld significant amounts of information, although several balked at answering one or two selected specific questions. The response ratio in terms of questions asked was well over 90 per cent.

As indicated in Table C.1, the survey covered recipients of 88 per cent of the exploring and producing companies' payments to contractors in local currency and 70 per cent of their payments to suppliers in local currency. Overall, of the petroleum industry's local payments to suppliers and contractors in 1965 amounting to £15,793,000, 84 per cent or £13,196,000 was received by firms covered by the survey. As was to be expected from the procedure of selecting firms to include in the survey, the great majority of firms omitted were small Nigerian-owned enterprises. This is shown by the fact that the survey included expatriate-owned firms receiving £12,301,000, which was 93 per cent of the total payments of £13,157,000 made by the petroleum industry to expatriate firms. On the other hand, the survey covered Nigerian firms receiving £895,000, which was only 34 per cent of the £2,636,000 paid locally in 1965 by the industry to Nigerian-owned firms.

The questionnaire employed verbally in the survey was roughly the following:

1. What were your gross receipts from petroleum exploring and producing companies in 1965?

2. What percentage were they of your total receipts?

3. Would you break down this total among the various petroleum exploring and producing companies?

4. How many Nigerians did you employ on the average during 1965?

5. What is your expected increment or decrement during 1966?

6. What were your total wages paid to Nigerians in 1965?

7. By how much have you increased your employment of Nigerians because of increased receipts from the petroleum producing or exploring companies since 1961?

8. How many expatriates did you employ in 1965?

9. By how much have you increased your expatriate employment because of increased receipts from petroleum companies since 1961?

10. What percentage of your gross receipts from petroleum companies was paid in company income taxes in 1965?

11. How much pay-as-you-earn (PAYE) withholding tax did you pay on behalf of your Nigerian employees in 1965?

12. How much PAYE did you pay on behalf of your expatriate employees in 1965? If none, could you estimate what these employees paid in income taxes during 1965?

13. What percentage of your gross receipts was paid abroad as imported materials, imported capital equipment, loan repayments, profit remittances, or remitted expatriate salaries?

14. What is the gross value of your total plant and equipment, and over what average period do you depreciate it?

15. What percentage of your gross receipts from the petroleum companies is spent on imported and local materials?

16. By how much have you expanded your total fixed assets because of increased petroleum business?

17. How did you finance this expansion?

18. What are your major sources of competition, Nigerian or expatriate firms?

19. What are your arrangements for payments from the petroleum companies? Are you paid wholly in local currency or is part of your payment made offshore? If so, what part is received in Nigeria?

The major problems in interviewee interpretation arose with questions 13, 15, and 16. The most sensitive question for Nigerian

firms was 10, though most companies did not pay any company taxes in 1965. For expatriate firms the most sensitive area concerned numbers of expatriates employed and salaries paid to them. This was probably a reflection of the difficult problems that the expatriate firms were then having in obtaining visas for expatriate personnel. In any event it was impossible to obtain reliable information regarding the salaries paid to expatriates. The expatriate tax data are also incomplete.

To move from the survey data included in Table C.1 to likely totals for all supplying and contracting firms, it is necessary to employ a blowup procedure so as to allow for firms omitted from the survey coverage. Since actual total local payments made by the exploring and producing companies to suppliers and contractors by category in 1965 were known (column 21 in Table C.1), these amounts could be applied to interviewed firms' receipts paid locally by the petroleum industry to arrive at blowup factors for each category (column 22). In effect, this implicitly assumes that the share of entries of noninterviewed supplier and contractor firms was proportional to their share of receipts from the exploring and producing companies. In view of the skewed distribution of noninterviewed firms, this assumption may not be particularly good, but no better alternative is available. To arrive at estimates of total activity in 1965 for the supplier or contractor firms, one need only apply the blowup factors to any or all of the columns 2 through 17. Though this procedure has faults, it should offer reasonably close orders of magnitude. Overall it is felt that a fairly high degree of confidence can be placed in most of the results of this survey of firms supplying or servicing the petroleum exploring and producing industry in Nigeria.

APPENDIX D

The Input-Output Data

TABLE D.1. 1965 A MATRIX AND TRANSACTIONS MATRIX

Inputs	Outputs			
	Petroleum[a]	Agriculture[c]	Manufacturing and construction[d]	Services[e]
Petroleum[b]	.0013	.0000	.0207	.0023
Agriculture[c]	.0000	.0000	.2202	.0004
Mfg. & construction[d]	.2056	.0011	.1581	.0715
Services[e]	.1205	.0024	.1712	.0574

(Million £ Nigerian)

	Petroleum[a]	Agriculture[c]	Manufacturing and construction[d]	Services[e]	Total intermediate use	Final demand	Total use
Petroleum[b]	.064	.000	8.271	1.064	9.399	39.846	49.245
Agriculture[c]	.000	.013	88.005	.198	88.216	716.764	804.980
Mfg. & construction[d]	10.126	.905	63.193	33.369	107.593	292.160	399.753
Services[e]	5.934	1.938	68.435	26.830	103.137	364.526	467.663
Total purchases	16.124	2.856	227.904	61.461	308.345		
Value added	33.121	802.124	171.849	406.202		1413.269	
Total production	49.245	804.980	399.753	467.663			1721.641

Data from Carter, Table 5.4; Chapter Seven, author's estimates.
[a] Coefficients and flows based on data in Table 5.4.
[b] Coefficients and flows based on Chapter Seven and author's estimates.
[c] Coefficients and flows based on aggregation of Carter's sectors 1, 2, 3, 6, and 7, adjusted from fiscal year 1959–60 to calendar year 1965.
[d] Coefficients and flows based on aggregation of Carter's sectors 4, 5, 10, 14, 16, 17, 18, and 19, adjusted from fiscal year 1959–60 to calendar year 1965.
[e] Coefficients and flows based on aggregation of Carter's sectors 8, 9, 11, 12, 13, 15, and 20 (excluding petroleum), adjusted from fiscal year 1959–60 to calendar year 1965.

TABLE D.2. *A* MATRICES, 1966–73

Inputs	Outputs			
	Petroleum	Agriculture	Manufacturing and construction	Services
1966				
Petroleum	.0016	.0000	.0207	.0023
Agriculture	.0000	.0000	.2202	.0004
Mfg. & construction	.2532	.0011	.1581	.0715
Services	.1487	.0024	.1712	.0574
1967				
Petroleum	.0014	.0000	.0207	.0023
Agriculture	.0000	.0000	.2202	.0004
Mfg. & construction	.2247	.0011	.1581	.0715
Services	.1320	.0024	.1712	.0574
1968				
Petroleum	.0013	.0000	.0207	.0023
Agriculture	.0000	.0000	.2202	.0004
Mfg. & construction	.2417	.0011	.1581	.0715
Services	.1419	.0024	.1712	.0574
1969, HIGH AND LOW PROJECTIONS				
Petroleum	.0006	.0000	.0207	.0023
Agriculture	.0000	.0000	.2202	.0004
Mfg. & construction	.1007	.0011	.1581	.0715
Services	.0591	.0024	.1712	.0574
1970, HIGH PROJECTION				
Petroleum	.0005	.0000	.0207	.0023
Agriculture	.0000	.0000	.2202	.0004
Mfg. & construction	.0821	.0011	.1581	.0715
Services	.0432	.0024	.1712	.0574
1971, HIGH PROJECTION				
Petroleum	.0005	.0000	.0207	.0023
Agriculture	.0000	.0000	.2202	.0004
Mfg. & construction	.0713	.0011	.1581	.0715
Services	.0418	.0024	.1712	.0574
1972, HIGH PROJECTION				
Petroleum	.0003	.0000	.0207	.0023
Agriculture	.0000	.0000	.2202	.0004
Mfg. & construction	.0543	.0011	.1581	.0715
Services	.0319	.0024	.1712	.0574

The Input-Output Data

TABLE D.2 (*continued*)

	Outputs			
Inputs	Petroleum	Agriculture	Manufacturing and construction	Services

1973, HIGH PROJECTION

Petroleum	.0003	.0000	.0207	.0023
Agriculture	.0000	.0000	.2202	.0004
Mfg. & construction	.0446	.0011	.1581	.0715
Services	.0262	.0024	.1712	.0574

1970, LOW PROJECTION

Petroleum	.0004	.0000	.0270	.0023
Agriculture	.0000	.0000	.2202	.0004
Mfg. & construction	.0812	.0011	.1581	.0715
Services	.0425	.0024	.1712	.0574

1971, LOW PROJECTION

Petroleum	.0004	.0000	.0270	.0023
Agriculture	.0000	.0000	.2202	.0004
Mfg. & construction	.0685	.0011	.1581	.0715
Services	.0401	.0024	.1712	.0574

1972, LOW PROJECTION

Petroleum	.0004	.0000	.0270	.0023
Agriculture	.0000	.0000	.2202	.0004
Mfg. & construction	.0551	.0011	.1581	.0715
Services	.0324	.0024	.1712	.0574

1973, LOW PROJECTION

Petroleum	.0003	.0000	.0270	.0023
Agriculture	.0000	.0000	.2202	.0004
Mfg. & construction	.0463	.0011	.1581	.0715
Services	.0273	.0024	.1712	.0574

Data are from Tables 5.5 and D.1 and author's estimates. Coefficients for petroleum are based on data in Table 5.5, adjusted by author's estimates. Coefficients for agriculture, manufacturing and construction, and services, from Table D.1, 1965 A Matrix, are assumed to remain constant.

TABLE D.3. $(I - A)^{-1}$ MATRICES, 1965–73

Outputs	Inputs			
	Petroleum	Agriculture	Manufacturing and construction	Services
1965				
Petroleum	1.007	0.000	0.026	0.004
Agriculture	0.058	1.000	0.264	0.021
Mfg. & construction	0.261	0.002	1.213	0.093
Services	0.176	0.003	0.224	1.078
1966				
Petroleum	1.009	0.000	0.026	0.004
Agriculture	0.071	1.000	0.268	0.021
Mfg. & construction	0.322	0.002	1.215	0.093
Services	0.218	0.003	0.225	1.079
1967				
Petroleum	1.008	0.000	0.026	0.004
Agriculture	0.063	1.000	0.267	0.021
Mfg. & construction	0.286	0.002	1.214	0.093
Services	0.193	0.003	0.225	1.078
1968				
Petroleum	1.008	0.000	0.026	0.004
Agriculture	0.068	1.000	0.268	0.021
Mfg. & construction	0.307	0.002	1.215	0.093
Services	0.208	0.003	0.225	1.078
1969, HIGH AND LOW PROJECTIONS				
Petroleum	1.003	0.000	0.026	0.004
Agriculture	0.028	1.000	0.264	0.021
Mfg. & construction	0.127	0.002	1.210	0.092
Services	0.086	0.003	0.222	1.078
1970, HIGH PROJECTION				
Petroleum	1.003	0.000	0.026	0.004
Agriculture	0.023	1.000	0.266	0.021
Mfg. & construction	0.104	0.002	1.209	0.092
Services	0.070	0.003	0.222	1.078
1971, HIGH PROJECTION				
Petroleum	1.003	0.000	0.026	0.004
Agriculture	0.019	1.000	0.266	0.021
Mfg. & construction	0.090	0.002	1.209	0.092
Services	0.061	0.003	0.221	1.078

The Input-Output Data

TABLE D.3 (*continued*)

Outputs	Inputs			
	Petroleum	Agriculture	Manufacturing and construction	Services
1972, HIGH PROJECTION				
Petroleum	1.002	0.000	0.026	0.004
Agriculture	0.015	1.000	0.266	0.021
Mfg. & construction	0.069	0.002	1.209	0.092
Services	0.046	0.003	0.221	1.078
1973, HIGH PROJECTION				
Petroleum	1.002	0.000	0.003	0.004
Agriculture	0.012	1.000	0.266	0.021
Mfg. & construction	0.056	0.002	1.208	0.092
Services	0.038	0.003	0.221	1.078
1970, LOW PROJECTION				
Petroleum	1.003	0.000	0.025	0.004
Agriculture	0.023	1.000	0.264	0.021
Mfg. & construction	0.102	0.002	1.198	0.091
Services	0.053	0.003	0.093	1.068
1971, LOW PROJECTION				
Petroleum	1.002	0.000	0.025	0.004
Agriculture	0.019	1.000	0.264	0.020
Mfg. & construction	0.086	0.002	1.198	0.091
Services	0.049	0.003	0.093	1.068
1972, LOW PROJECTION				
Petroleum	1.002	0.000	0.025	0.004
Agriculture	0.015	1.000	0.264	0.020
Mfg. & construction	0.069	0.002	1.198	0.091
Services	0.040	0.003	0.092	1.068
1973, LOW PROJECTION				
Petroleum	1.002	0.000	0.025	0.004
Agriculture	0.013	1.000	0.264	0.020
Mfg. & construction	0.058	0.002	1.197	0.091
Services	0.033	0.003	0.092	1.068

Data are from Table D.1 and corresponding matrices of Table D.2.

TABLE D.4. 1965–68 Y VECTORS
(*Million £ Nigerian*)

Item	1965	1966	1967	1968
Petroleum	32	44	51	18
Agriculture, fishing, and forestry	802	811	819	826
Manufacturing, construction, and nonpetroleum mining	172	180	183	193
Services, including all others	406	428	432	444
Total[a]	1,414	1,463	1,485	1,481

Data from Appendix Table A.1; author's estimates. Value added to Gross Domestic Product at market prices.

[a] Individual components may not sum to total because of rounding.

TABLE D.5. 1965–68 X VECTORS
(*Million £ Nigerian*)

Item	1965	1966	1967	1968
Petroleum	49	74	79	29
Agriculture, fishing, and forestry	805	814	822	829
Manufacturing, construction, and nonpetroleum mining	400	419	426	449
Services, including all others	468	493	497	511
Total	1,722	1,800	1,824	1,818

Data from Tables D.1, D.2, and D.4. Gross output.

The Input-Output Data 201

TABLE D.6. 1965–68 T VECTORS
(Million £ Nigerian)

Item	1965	1966	1967	1968
Foreign exchange:				
Petroleum[a]	36	58	51	48
Nonpetroleum[b]	291	252	221	240
Total	327	310	272	288
Investment resources:				
Petroleum[c]	50	65	70	50
Nonpetroleum[d]	184	178	167	181
Total	234	243	237	231
Skilled labor:				
Petroleum[e]	2	3	2	1
Nonpetroleum[f]	102	108	114	120
Total	104	111	116	121

Data from Tables 5.2, 5.5, 6.6, A.1, and A.2; *Nigerian Human Resource Development and Utilization*, pp. 18–26; Charles R. Frank, Jr., "Employment and Economic Growth in Nigeria," (Washington, D.C.: U.S. Agency for International Development, 1966); Nigeria [9]; and author's estimates.

[a] Imports of goods and services from Table 5.5.
[b] Total nonpetroleum foreign exchange use from Table 6.6.
[c] Unpublished estimate from aggregation of individual company data.
[d] Residuals from subtraction of petroleum entries from gross fixed capital formation from Table A.1. Note that figures for 1965 and 1966 are from *Economic Indicators, December 1968* (Lagos: Federal Office of Statistics, 1969), p. 37, and figures for 1967 and 1968 are author's estimates.
[e] Estimated as two-thirds of local salaries and wages from Table 5.2.
[f] Estimates based on information contained in *Nigerian Human Resource Development and Utilization*; Frank, "Employment and Economic Growth," and Nigeria [9]. Adjustments of Frank's figures for 1963/64 indicate that perhaps 282,000 skilled laborers were employed in 1965. To find a reasonable growth rate of skilled labor, the following reasoning was used (from *Nigerian Human Resource Development and Utilization*): of total 1965 working force of about 17.5 million people (p. 18), 282,000 or 1.6 per cent were skilled; annual labor force attrition might be 350,000 (p. 19), of which possibly 1.6 per cent or about 6,000 are skilled laborers; total labor force might grow at 2.5 per cent (p. 20) or 440,000 from 1965 level, of which about 55,000 have post-primary education (p. 20); annual addition of skilled laborers is assumed to amount to 40 per cent of post-primary educated new entrants, or 22,000 in 1966; net addition of skilled laborers is thus 16,000, or 22,000 less 6,000; the implied rate of growth is 5.75 per cent per annum. Skilled labor is thus assumed to grow at 5.75 per cent annually from a 1965 base of 282,000. To measure skilled labor use in financial terms, an average salary of £360 per year is used throughout (Nigeria [9]).

TABLE D.7. *F* MATRICES, 1965–73

Inputs	Outputs			
	Petroleum	Agriculture	Manufacturing and construction	Services
		1965		
Foreign exchange	0.731	0.141	0.250	0.163
Investment resources	1.015	0.025	0.252	0.144
Skilled labor	.047	0.002	0.052	0.184
		1966		
Foreign exchange	0.784	0.097	0.251	0.136
Investment resources	0.878	0.023	0.228	0.130
Skilled labor	0.041	0.002	0.053	0.184
		1967		
Foreign exchange	0.646	0.085	0.216	0.119
Investment resources	0.886	0.022	0.209	0.121
Skilled labor	0.025	0.002	0.052	0.181
		1968		
Foreign exchange	1.655	0.092	0.223	0.125
Investment resources	1.724	0.023	0.216	0.127
Skilled labor	0.034	0.002	0.051	0.188
	1969, HIGH AND LOW PROJECTIONS			
Foreign exchange	1.154	0.119	0.250	0.149
Investment resources	0.686	0.023	0.228	0.130
Skilled labor	0.024	0.002	0.053	0.184
	1970, HIGH PROJECTION			
Foreign exchange	0.835	0.119	0.250	0.149
Investment resources	0.409	0.023	0.228	0.130
Skilled labor	0.011	0.002	0.053	0.184
	1971, HIGH PROJECTION			
Foreign exchange	0.687	0.119	0.250	0.149
Investment resources	0.355	0.023	0.228	0.130
Skilled labor	0.008	0.002	0.053	0.184
	1972, HIGH PROJECTION			
Foreign exchange	0.652	0.119	0.250	0.149
Investment resources	0.219	0.023	0.228	0.130
Skilled labor	0.006	0.002	0.053	0.184

TABLE D.7 (*continued*)

Inputs	Outputs			
	Petroleum	Agriculture	Manufacturing and construction	Services

1973, HIGH PROJECTION

Foreign exchange	0.615	0.119	0.250	0.149
Investment resources	0.181	0.023	0.228	0.131
Skilled labor	0.005	0.002	0.053	0.184

1970, LOW PROJECTION

Foreign exchange	0.858	0.119	0.250	0.149
Investment resources	0.438	0.023	0.228	0.130
Skilled labor	0.012	0.002	0.053	0.184

1971, LOW PROJECTION

Foreign exchange	0.687	0.119	0.250	0.149
Investment resources	0.374	0.023	0.228	0.130
Skilled labor	0.010	0.002	0.053	0.184

1972, LOW PROJECTION

Foreign exchange	0.657	0.119	0.250	0.149
Investment resources	0.275	0.023	0.228	0.130
Skilled labor	0.008	0.002	0.053	0.184

1973, LOW PROJECTION

Foreign exchange	0.623	0.119	0.250	0.149
Investment resources	0.232	0.023	0.228	0.130
Skilled labor	0.007	0.002	0.053	0.184

Data for the first four matrices are from Tables D.5 and D.6 and author's estimate; data for the remaining matrices are from the 1965 and 1966 F matrices and author's estimate.

Petroleum coefficients for the first four matrices are estimated as the ratio of petroleum factor uses from Table D.6 to petroleum gross output from Table D.5. The remainder are estimated from projections of the petroleum industry's requirements.

The other coefficients from the first four matrices are estimated from the computer iterations that solve for F in the equation

$$\underset{3 \cdot 1}{T} = \underset{3 \cdot 4}{F} \cdot \underset{4 \cdot 1}{X}$$

with data taken from Tables D.5 and D.6. In the remaining matrices, the foreign exchange figures for agriculture, manufacturing and construction, and services are averages of coefficients for 1965 and 1966. The figures for investment resources and skilled labor are coefficients for 1966.

TABLE D.8. 1969–73 U VECTORS
(Million £ Nigerian)

Item	1969	1970	1971	1972	1973
		HIGH PROJECTION			
Foreign exchange:					
Petroleum[a]	198	353	440	544	644
Nonpetroleum[b]	240	250	265	275	285
Total	438	603	705	829	929
Investment resources:					
Petroleum[c]	108	195	274	291	345
Nonpetroleum[d]	176	177	188	199	207
Total	284	372	462	490	552
Skilled labor:					
Petroleum[e]	3	3	3	3	3
Nonpetroleum[f]	127	135	143	151	159
Total	130	138	146	154	162
		LOW PROJECTION			
Foreign exchange:					
Petroleum[a]	198	306	351	417	484
Nonpetroleum[b]	240	250	265	275	285
Total	438	556	616	692	769
Investment resources:					
Petroleum[c]	103	173	226	244	276
Nonpetroleum[d]	176	178	179	200	207
Total	279	351	405	444	483
Skilled labor:					
Petroleum[e]	3	3	3	3	3
Nonpetroleum[f]	127	135	143	151	159
Total	130	138	146	154	162

Data from Tables 8.2, 8.3, 8.5, and 8.7; *Nigerian Human Resource Development and Utilization*, pp. 18–26; Charles R. Frank, Jr., "Employment and Economic Growth in Nigeria" (Washington, D.C.: U.S. Agency for International Development, 1966); Nigeria [9]; and author's estimates.

[a] Sum of petroleum balance-of-payments impact plus imports of materials plus imports of services plus net factor income paid abroad from Table 8.5.

[b] Nonpetroleum foreign exchange availability from Table 8.7.

[c] Sum of petroleum industry payments to government from Table 8.3 plus author's estimates of the petroleum industry's own investment.

[d] Estimated as 12 per cent of nonpetroleum Gross Domestic Product in the previous year from Tables 8.11 and 8.12; percentage based on 1962–66 average.

[e] Estimated as two-thirds of local wages and salaries from Table 8.2.

[f] Author's estimates, assuming an annual growth rate for the supply of skilled laborers of 5.75 per cent from a 1968 level of 332,000 and a constant average salary rate of £360 per man. For explanation see note *f* to Table D.6.

APPENDIX E

The Petroleum Industry's Value Added and Balance-of-Payments Impact

The purpose of this appendix is to demonstrate algebraically the calculations of the petroleum industry's value added, employing both domestic and national product accounting, and of the industry's balance-of-payments impact. In addition, a comparison of the two concepts is undertaken. Finally, a comparison is made of value added under the two assumptions that output is calculated first at posted prices and then at realization prices. Symbols used in the analysis are as follows:

R = receipts
X = export proceeds
L = local proceeds
F = foreign capital inflows
E = expenditures
M_g = imports of goods
M_s = imports of services
G = payments to government
N_o = other net factor payments abroad
N = net factor payments abroad
B = balance-of-payments impact
V_d = value added using domestic accounting concepts
V_N = value added using national accounting concepts
V_d^P = value added, domestic, at posted prices
V_d^R = value added, domestic, at realized prices
X^P = export proceeds at posted prices
X^R = export proceeds at realized prices
V_N^P = value added, national, at posted prices
V_N^R = value added, national, at realized prices

H = harbor dues
W = local wages and salaries
L_g = local goods
L_s = local services
D = depreciation
P = profits

Definitional Relationships
(1) $R \equiv X + L + F$
(2) $E \equiv M_g + M_s + G + H + W + L_g + L_s + D + P + N_o$
(3) $R \equiv E$
(4) $N \equiv D + P + N_o$

Balance-of-Payments Impact
(5) $\begin{aligned}B &= X + F - M_g - M_s - D - P - N_o \\ &= G + H + W + L_g + L_s - L\end{aligned}$

Note that the calculation of B is insensitive to the price used for valuation of export proceeds.

Value Added

Value added as a residual:

(6) $\begin{aligned}V_d &= R - M_g - M_s - L_g - L_s \\ &= X + L + F - M_g - M_s - L_g - L_s\end{aligned}$

Value added built up from individual components:

(7) $\begin{aligned}V_d &= E - M_g - M_s - L_g - L_s \\ &= G + H + W + D + P + N_o\end{aligned}$

It is clear from equations (1)–(3) that

(8) $P = X + L + F - G - H - W - L_g - L_s - D - M_g - M_s - N_o$

Substituting (8) in (7),

(9) $\begin{aligned}V_d &= G + H + W + D + N_o + X + L + F - G - H - W - D - N_o - M_g - M_s - L_g - L_s \\ &= X + L + F - M_g - M_s - L_g - L_s\end{aligned}$

Value added built up from individual components:

By definition,

(10) $\qquad V_N = V_d - N = V_d - D - P - N_o$

Substituting (7) in (10),

(11) $\qquad\qquad V_N = G + H + W$

Value Added and Balance-of-Payments Impact

Value added as a residual:

(12) $V_N = X + L + F - M_g - M_s - L_g - L_s - D - P - N_o$

Substituting (8) in (12),

(13) $V_N = X + L + F - M_g - M_s - L_g - L_s - D - N_o - X - L - F + M_g + M_s + L_g + L_s + D + N_o + G + H + W$
$= G + H + W$

Note that two counteracting biases are introduced: (1) All imports of services are excluded from value added. But some portion should be included to the extent that this item actually consists of offshore payments to contractors and these contractors then make factor payments that are not included elsewhere. (2) All harbor dues are included as value added. But some portion should be excluded, perhaps one-third, because services are actually rendered to this extent.

Comparison of Balance-of-Payments Impact and Value Added

Comparing (5) and (6) or (5) and (7),

(14) $\qquad V_d = B - N - L + L_g + L_s$

Thus if $(N + L) < (L_g + L_s)$, then $B > V_d$ by the same amount; and conversely, if $(N + L) > (L_g + L_s)$, then $B < V_d$ by the same amount.

Comparing (5) and (11) or (5) and (12), or substituting (10) in (14),

(15) $\qquad V_N = B - L + L_g + L_s$

Thus if $(L) < (L_g + L_s)$, then $B > V_N$ by the same amount; and conversely, if $(L) > (L_g + L_s)$, then $B < V_N$ by the same amount.

Comparison of Value Added Calculated at Posted Prices and at Realization Prices

Since all other variables in (6) and (7) are unchanged,

(16) $\qquad (V_d^P > V_d^R) = (X^P > X^R)$

Since export variables cancel in the residual calculation (see [12] and [13]) and do not enter the components calculation (see [11]),

(17) $\qquad V_N^P = V_N^R$

Notes

Chapter One

Full authors' names, titles, and publication data for works cited in the Notes will be found in the Selected Bibliography, pp. 221–26. Government publications are identified in the Notes and Bibliography by bracketed numbers and are listed in the bibliography under Nigeria.

1. This chapter in part relies on Gabriel, pp. 38–69 and 252–55, for both data and interpretation. Major analyses of the international petroleum industry drawn on include Penrose, Frankel, Adelman, and Hartshorn.

2. *Statistical Review of the World Oil Industry 1968* (London: The British Petroleum Company Limited, 1969), p. 21.

3. *Ibid.*, p. 20.

4. *Energy Policy*, p. 46.

5. S. L. McDonald, *Federal Tax Treatment of Income from Oil and Gas* (Washington, D.C.: The Brookings Institution, 1963), p. 32, cited in Gabriel, p. 47.

6. *Fortune*'s 1969 ranking of the largest American industrial organizations placed the five U.S.-controlled major petroleum companies (Standard Oil [N.J.], Mobil, Texaco, Gulf, and Standard Oil of California) in the 2nd, 7th, 8th, 9th, and 16th rankings. The two foreign-controlled international majors (Royal Dutch Shell, and British Petroleum) were 1st and 3d among *Fortune*'s foreign firms in 1968. Twentieth spot in the foreign ranking was occupied by Compagnie Française des Pétroles, a French-controlled firm that is occasionally included as a major. In addition, several other petroleum companies, the American or foreign independents, occupy prominent positions in these rankings. See "The Fortune Directory of the 500 Largest Industrial Corporations," *Fortune*, LXXVII (May 15, 1969), p. 168, and "The Fortune Directory of the 200 Largest Industries Outside the U.S., *Fortune*, LXXVI (September 15, 1968), p. 131.

7. The system of crude prices evolved from a U.S. Gulf Coast basing-point system to a London equalization in the late 1940's and eventually

to a New York equalization in the 1950's. Posted prices in the producing countries were established to equal the delivered price at the equalization point less transportation from the source. For one example of the many discussions of this topic see Hartshorn, pp. 127–65.

8. This discussion modifies Gabriel's analysis of this subject. See Gabriel, pp. 55–56.

9. During this period, U.S. tax legislation allowed petroleum companies to deduct $27\frac{1}{2}$ per cent of gross income from taxable income on a given property with the deduction limited to 50 per cent of net income on the property. In practice, depletion allowances have turned out to be unimportant in the Eastern Hemisphere operations of the American majors.

10. The 1965 figure was made up of 7 majors, 34 U.S. independents, 31 state-owned companies based in producing areas, 14 companies based in producing areas, and 18 third country independents. See Gabriel, p. 66.

11. A concise summary of the views of the export pessimists is contained in Gerald M. Meier, "The Carry-over Problem," in Gerald M. Meier (ed.), *Leading Issues in Development Economics* (New York: Oxford University Press, 1964), pp. 371–76. Original statements include, among others, Hans W. Singer, "The Distribution of Gains between Investing and Borrowing Countries," *The American Economic Review*, XL (May 1950), pp. 473–85; Gunnar Myrdal, *An International Economy* (New York: Harper and Row, 1956), especially pp. 99–118; Ragnar Nurkse, "International Investment To-day in the Light of Nineteenth-Century Experience," *The Economic Journal*, LXIV (December 1954), pp. 744–58; and Raul Prebisch, "Commercial Policy in the Underdeveloped Countries," *The American Economic Review*, XLIX (May 1959), pp. 251–73.

12. In 1960 five of the major producing countries, Venezuela, Iran, Iraq, Kuwait, and Saudi Arabia, formed the Organization of Petroleum Exporting Countries (OPEC), which has evolved into a loosely organized producers' intergovernmental organization that coordinates oil policy. Since its formation, Libya, Indonesia, Qatar, Abu Dhabi, and Algeria have also joined OPEC.

13. But useful cross-country theorizing and analysis might very well precede the undertaking of individual country studies.

Chapter Two

1. For a discussion of the main geological features of Nigerian oil, see Frankl and Cordry, pp. 1–6.

2. For a related discussion applying to the petroleum industry in Venezuela, see Gabriel, pp. 244–48.

3. Dickie, p. 1.

4. *The Shell-BP Story*, p. 8.

5. *Ibid.*, pp. 8–11. Under Nigerian law, Shell-BP had to give up at least

half of its OPL acreage upon conversion of its concession into OML's. For legal details see Appendix B.

6. Frankl and Cordry, p. 7.
7. Dickie, p. 2.
8. "Nigeria Jumps Tax Ante on Oil." The following discussion is based on this account.

Chapter Three

1. In the past four years alone eight major studies of the Nigerian economy have appeared: Aboyade, *Foundations of an African Economy*; Clark, "The Choice of Optimal Import Substitution Patterns for Nigeria"; Diejomaoh, "Financing Development Expenditures: Nigerian Experience since 1950"; Helleiner, *Peasant Agriculture, Government, and Economic Growth in Nigeria*; Kilby, *Industrialization in an Open Economy: Nigeria 1945–1966*; Lewis, *Reflections on Nigeria's Economic Growth*; Oluwasanmi, *Agriculture and Nigerian Economic Development*; and Stolper, *Planning Without Facts: Lessons in Resource Allocation from Nigeria's Development*.
2. Helleiner, *Peasant Agriculture*, p. 5.
3. *Ibid.*, p. 18.
4. Nigeria, Federal Office of Statistics, *Economic Indicators*, IV (December 1968), p. 35. There is an irreconcilable break in accounting procedures between 1957 and 1958.
5. Table A.1 does not incorporate the changes in the national income accounts carried out by the Federal Office of Statistics in 1968. These changes are based on the inflated but official 1963 census results and hence offer no additional insights into recent Nigerian growth patterns. Unless otherwise specified, figures from Table A.1 are employed in this study. For a comprehensive discussion of the vagaries associated with Nigeria's national income accounts, see Clark, pp. 35–100. Lewis (pp. 9–13) qualifies the official data and arrives at an annual growth rate for GDP of 4.4 per cent from 1950 to 1963.
6. Bruce F. Johnston, "Agriculture's Role," p. 4.
7. See Helleiner, *Peasant Agriculture*, p. xii, and Lewis, p. 15.
8. Helleiner, *Peasant Agriculture*, p. 50, points out that export diversification at the national level is helpful if (1) all export prices do not move in unison, and (2) there is some mechanism for transferring resources from the temporarily more prosperous regions to the temporarily depressed ones.
9. Lewis, p. 20.
10. Helleiner, *Peasant Agriculture*, p. 28.
11. *Ibid.*, p. 134. See also Lewis, p. 20.

12. For an early discussion of manufacturing in Nigeria, see Sokolski, especially pp. 229–86.

13. Kilby, *Industrialization*, pp. 81–181; see also Kilby's introduction to the Nigerian industrial sector, pp. 16–22, and his concluding chapter on industrial strategy, pp. 345–64.

14. *Ibid.*, p. 17.

15. Lewis, pp. 25–26.

16. An important exception is Kilby, *The Development of Small Industry in Eastern Nigeria*. For a summary, see Kilby, *Industrialization*, p. 18.

17. John R. Harris has carried out extensive analyses on Nigerian entrepreneurs; see especially his "Factors Affecting the Supply of Industrial Entrepreneurship in Nigeria" (Massachusetts Institute of Technology, Cambridge, Massachusetts, December 1966).

18. For a related discussion, see Kilby, *Industrialization*, pp. 22–25.

19. See Helleiner, *Peasant Agriculture*, pp. 301–2.

20. Lewis, pp. 48–53, and Helleiner, *Peasant Agriculture*, pp. 302–5.

21. Helleiner, *Peasant Agriculture*, pp. 305–10. See especially *Nigerian Human Resource Development and Utilization*.

22. An optimistic view of the direct role of the Nigerian government is contained in Helleiner, *Peasant Agriculture*, pp. 311–20. For a discussion of some aspects on the other side of the ledger, see Pearson, "The Political Economics of Nigerian Short-Term Borrowing," pp. 337–60. Kilby (*Industrialization*, p. 23) has summarized the government's industrialization policy: "Measures taken by public authorities to promote industrialization are of two kinds, direct and indirect. Indirect measures include the provision of social infrastructure, guarantees to private investors against uncompensated nationalization, and freedom for foreigners regarding the sale of their assets and repatriation of profits. The direct measures may be grouped into three categories: fiscal incentives, support activities, and direct public investment in manufacturing."

23. Helleiner, *Peasant Agriculture*, p. 320.

24. Lewis, p. 5.

25. Lewis, especially pp. 21–22, where he advocates an elimination of taxation of agricultural exports; Bruce F. Johnston, "Agriculture's Role," pp. 38–41; and Glenn L. Johnson *et al.*

26. Lewis, p. 20.

27. See Lewis, p. 53, Bruce F. Johnston, "Agriculture's Role," pp. 40–41, and Helleiner, *Peasant Agriculture*, pp. 58–75. For a comparative study of peasant price response, see Walter P. Falcon, "Farmer Response to Price in a Subsistence Economy: The Case of West Pakistan," *The American Economic Review*, LIV (May 1964), 580–91.

28. Lewis, p. 42, states that between 1950–52 and 1961–63, the overall index for southern Nigerian prices paid to farmers fell from 100 to 73,

while that for wages paid by the federal government to unskilled labor increased from 100 to 297.

29. See Archibald C. Callaway, "Unemployment Among African School Leavers," *The Journal of Modern African Studies*, I (September 1963), 351–71.

30. Kilby (*Industrialization*, pp. 234–41) argues that underinvestment in secondary education has created a negative or zero marginal product for investment in primary education. For a detailed analysis of political aspects of policies for educational expansion in southern Nigeria, see Abernethy.

31. Lewis, p. 45, develops this theory for a solution to urban unemployment in the long run.

Chapter Four

1. For a preliminary report on a much broader study, see Raymond Vernon, "Multinational Enterprise and the Nation State," *Journal of Common Market Studies*, VIII (December 1969), 160–70.

2. Chenery and Strout, pp. II-7 to II-11.

3. A curious dichotomy is observed in the discussion of externalities in the literature of economics. Public finance theorists dealing principally with cost-benefit analysis in an advanced economy have tended to play down the importance of these phenomena, whereas those dealing with an underdeveloped economy have increasingly stressed their role. For an evaluation of this dichotomy see Hirschman, *Development Projects Observed*, pp. 174–85.

4. Backward and forward linkages were introduced by Hirschman in *The Strategy of Economic Development*, pp. 98–119. The term final demand linkages was coined by Watkins, in "A Staple Theory of Economic Growth."

5. Baldwin, pp. 64–70.

6. The originator of input-output analysis, Wassily W. Leontief, applies some of his ideas to less developed nations in "The Structure of Development," *Scientific American*, CCIX (September 1963), 148–54.

7. Jonathan V. Levin discusses how the absence of favorable conditions for final demand linkages can result in the creation of luxury importers in *The Export Economies: Their Pattern of Development in Historical Perspective* (Cambridge: Harvard University Press, 1960), p. 7ff.

8. Gerald M. Meier, "The Carry-Over Problem," in Gerald M. Meier, ed., *Leading Issues in Development Economics* (New York: Oxford University Press, 1964), p. 374.

9. See Alan S. Manne and Harry M. Markowitz, "Alternate Methods of Analysis," and H. M. Markowitz and Alan J. Rowe, "Metalworking Requirements Analysis," in Manne and Markowitz, pp. 8–19, 285–86, 293.

10. For a standard treatment of input-output theory, see Chenery and Clark, pp. 13–80.

11. For elaboration of these limitations in general see Robert Dorfman, "The Nature and Significance of Input-Output," *Review of Economics and Statistics*, XXXVI (May 1954), pp. 121–33.

12. For a detailed theoretical discussion of input-output, see Dorfman, Samuelson, and Solow, pp. 204–300.

13. The derivation of the basic equation of the model is summarized in the following six equations:

(1) $X = AX + Y$, by definition;
(2) $X - AX = Y$;
(3) $(I - A)X = Y$;
(4) $X = (I - A)^{-1} \cdot Y$;
(5) $T = FX$, by definition;

and, by substituting (4) into (5),

(6) $T = F \cdot (I - A)^{-1} \cdot Y$;

where $X \equiv n \times 1$ vector of n sectoral gross outputs, comprising both intermediate and final outputs, and other variables are as defined in the text.

14. Backward-linkage effects associated with capital goods industries must, of course, be handled separately.

Chapter Five

1. Manufacturing gross output as reported by the Industrial Surveys was the following for the years 1963, 1964, and 1965 (all figures in thousands of £ Nigerian): 136,623, 179,389, and 218,199. Wages and salaries for the same three years were 13,317, 16,520, and 20,456. At the same time total value added was 54,853, 68,733, and 85,827. Industrial costs are defined as gross output less value added; costs of materials were about 75 per cent of industrial costs. For 1963 data see Nigeria [10], p. 42. Data for 1964 are contained in the 1966 issue of this same *Abstract* (1968), p. 49. The 1965 data were obtained directly from the Federal Office of Statistics. Comparable 1966 data are available but only for the Northern and Western Regions and Lagos Federal Territory.

2. Nigerian Institute of Social and Economic Research, "Economic and Social Survey, 1958 to 1968" (Ibadan, Nigeria: Conference on National Reconstruction and Development in Nigeria, March 1969), p. 38.

Chapter Six

1. Reynolds (pp. 273–87) constructs a domestic expenditures model for the estimation of what he terms "returned value," i.e. payments for local goods and services plus local taxes. To investigate the effects of export price changes he extends his analysis to derive "returned value terms of trade" equal to the index of "returned value" per unit of the export divided

Notes to Pages 79–99

by the index of the unit price of imports. Stable petroleum prices make this extension unnecessary in discussing Nigerian oil.

2. The growth of imports should be viewed as lending inducement for future domestic production and hence potential linkages. See Hirschman, *Strategy*, pp. 120–25.

3. This figure is based on information regarding domestic intermediate inputs derived from the survey of suppliers and contractors. See Table C.1 in Appendix C.

4. This discussion is based on Roemer, pp. 265–69. Employing Roemer's notation, the equation for the import content of third and all subsequent rounds can be derived algebraically. For purposes of this analysis it is necessary to make the unrealistic assumptions that m' and d apply to all sectors of the Nigerian economy. The following notation is employed:

S = value of goods and services supplied locally to the petroleum industry;
d = fraction of S that is domestic intermediate inputs;
S' = value of domestic intermediate inputs purchased to produce S;
S'' = value of domestic intermediate inputs purchased to produce S', etc.;
M' = import content of second and all subsequent rounds;
m' = fraction of S that is import content; and
M'' = import content of third and all subsequent rounds.

Using this notation, the expression for M'' is derived in the following manner:

(1) $S' = dS$; $S'' = ddS = d^2S$; $S''' = d^3S$; etc., by definition.
(2) $M' = m'(S + S' + S'' + S''' + \ldots)$, by definition. Hence
(3) $M' = m'S(1 + d + d^2 + d^3 + \ldots)$, and thus
(4) $M' = m'S/(1-d)$.
(5) $M'' = M' - m'S$ by definition.
(6) $M'' = m'dS/(1-d)$, by substitution of (4) into (5).

5. This discussion applies specifically to Shell-BP and in general to most of the other companies whose Nigerian operations have been of shorter duration and of more limited scope.

6. Nigeria [3], p. 7, and [6], p. 5.

7. *The Shell-BP Story*, p. 33.

Chapter Seven

1. This information is contained in a press release issued by the Federal Ministry of Information, November 27, 1969.

2. Nigeria [3], p. 4.

3. Nigeria [11], p. 22. For the undercount estimate see Charles R. Frank, Jr., "Employment and Economic Growth in Nigeria," (Washington, D.C.: U.S. Agency for International Development, 1966), pp. 10–13.

4. Nigeria [9], p. 33.

5. These figures are taken from unpublished Federal Office of Statistics data pertaining to the 1965 industrial survey. The petroleum industry employed more people and paid more wages and salaries than any single manufacturing industry. The leading manufacturer was motor vehicle repairs, with 11,514 employees to whom it paid £3,343,000 in 1965.

Chapter Eight

1. The high projection is echoed in a forecast by *The Economist* that places Nigerian petroleum output at 2.5 million barrels per day in 1975. See "What Happens Next Time?," *The Economist* (June 10, 1967), p. 1134.

2. For comparative data on oil industry local payments in Venezuela, see Gabriel, pp. 170–79 and 184–86.

3. These rates are based on recent historical experience. See also Lewis, pp. 10–13.

4. Value added in sector j, V_j, is equal to
$$(1 - \sum_i a_{ij})$$
times gross output in sector j, X_j. This can be shown as follows. By definition,
$$X_j = V_j + \sum_i X_{ij},$$
where X_{ij} is the total amount of output of sector i purchased by sector j. And also, $a_{ij}X_j = X_{ij}$, where a_{ij} is a Leontief coefficient. Hence, $\sum_i a_{ij}X_j = \sum_i X_{ij}$, and therefore $X_j = V_j + \sum_i a_{ij}X_j$. The stated relationship is found by rearranging and collecting terms.

5. See Carter and Table D.1.

6. Lewis, p. 58.

7. *Nigerian Human Resource Development and Utilization*, pp. 18–26. Data on wages paid per sector are found in Nigeria [9].

8. These relationships can be stated in terms of the notation employed earlier, letting the subscript 1 refer to the petroleum sector and assuming four sectors in the economy:

(1) $X_1 = r_{11}Y_1 + r_{12}Y_2 + r_{13}Y_3 + r_{14}Y_4$

(2) $\sum_{i=1}^{4} X_{i1} = X_1 - V_1$

(3) $\sum_{i=2}^{4} X_{i1} = X_1 - V_1 - X_{11}$

where
X_1 = gross output of the petroleum sector;
r_{ij} = the element of the ith row and jth column of the 4×4 Leontief inverse matrix $(I - A)^{-1}$;
Y_i = deliveries to final demand of sector i, $i = 1 \ldots 4$;

Notes to Pages 134–39

X_{ti} = amount of the output of sector i used by the petroleum sector;
V_1 = value added of the petroleum sector.

9. For a related discussion, see Sokolski, pp. 119–21.

10. See "Nigeria's Oil Revolution Begins," p. 235.

11. In a talk before the American-Nigerian Chamber of Commerce on December 14, 1967, the Permanent Secretary of the Federal Ministry of Industries, Philip C. Asiodu, called for petroleum industry investment in oil-allied industries, especially petrochemicals, fertilizer production, and blending plants.

12. Asiodu, p. 12.

13. S. L. Clement and W. C. Scott, *Survey of Fertilizer Use in Nigeria: An Evaluation of Potential Demand and Methods of Supply* (Wilson Dam, Alabama: Tennessee Valley Authority, 1965), pp. 15–16.

14. *Ibid.*, pp. 16–17.

15. There is an additional speculative forward linkage that is fascinating to contemplate but unlikely to be important in the near future. This is the production of protein from petroleum and natural gas, a process for which technology is now being developed. See "Food from Petroleum," *SPAN*, Fall 1968, pp. 6–9.

Chapter Nine

1. For a discussion of politics in Nigeria that leads up to the civil war, see Schwarz, *Nigeria*, pp. 152–301.

2. "Already during the recent election crisis there were rumours, some even voiced by public figures, that the presence of oil in some parts of Nigeria and its absence in others was a factor threatening the unity of the Federation." "Nigeria's Oil Revolution Begins," *West Africa* (February 27, 1965), p. 233. For an earlier discussion of Nigerian petroleum, see Sokolski, pp. 41–45, 54–62.

3. The actual roles and attitudes of the petroleum companies in the civil war are matters of some dispute. In an article written just after the end of the war, Walter Schwarz, an authority on Nigerian political affairs, summed up the situation as follows: "Just after secession, another British Minister of State went to Biafra and reported after consultation with the oil companies there that *force majeure* had to be recognized and £20 million in oil royalties would have to be paid to the powers that be in Biafra as [Colonel] Ojukwu had demanded. But when the Minister got home, he found that the oil companies' decision had already been countermanded by the [British] government, which blocked payment of the money and used its majority holding in Shell-BP to veto the decision. The likely reason was that in the wake of the disruption of oil supplies from the Middle East after the Six-Day War, Mr. Wilson's government dared not run any risk with the Nigerian supplies.... The oil companies were

alarmed and apprehensive about secession ... [and] became active supporters of the federal cause—especially when new discoveries changed the emphasis from the East to the Mid-West and off-shore sources." Schwarz, "Foreign Powers and the Nigerian War," pp. 12–14.

4. A close observer of the Nigerian political scene, James O'Connell, is of the opinion that "The Ibo bureaucratic and military elites were alienated by the July [1966] coup and a majority of them took a secessionist stand. Initially they hoped to bring the Mid-West with them into secession. They reckoned that the resources in land and oil that the two regions possessed between them would be adequate for whatever economic problems arose." O'Connell, p. 177.

5. See Nigeria [20] and Felix M. Ani, "Petroleum and the Nigerian Crisis," reprint of a statement that was carried on several radio broadcasts and in various Nigerian newspapers during June 1967.

6. Evidence of this can be seen from an official Biafran publication: "Biafra's oil which was fetching that dead federation millions of pounds in foreign exchange every year is no more within Nigeria's reach" (Nigeria's War of Financial Suicide," *Biafra Newsletter*, I (January 12, 1968), p. 5). And from a different point of view, O'Connell sums it up this way: "[The Ibos] are basing their economic future on the huge oil deposits of the Niger Delta ... which are found for the most part in land belonging to the minorities who seethe with resentment at the small advantages that have come their way from the oil found on their land" ("Nigeria: The Politics of Majorities and Minorities," *Nigerian Opinion*, III (July 1967), 218).

Chapter Ten

1. Stolper, *Planning Without Facts*, p. 312.

2. Raymond Vernon, a leading analyst of relations between private foreign investors and host country governments, stresses the continual struggle over control and the difficulty that supposed enclave industries have in retaining separateness even if they so desire ("Foreign-owned Enterprise," pp. 361–80).

3. For discussion on this issue, see Schwarz, "Foreign Powers."

4. For an analysis of planning by the chief architect of Nigeria's first development plan, see Stolper, *Planning Without Facts*, especially pp. 26–51.

5. For further analysis of this question, see Carl K. Eicher and Glenn L. Johnson, "Policy for Nigerian Agricultural Development in the 1970's," in Eicher and Liedholm, pp. 376–92.

Selected Bibliography

Selected Bibliography

This bibliography is a selected list of works and documents chosen on the basis of immediate relevance to this study. A complete bibliography is available upon request from the author. Publications of the Nigerian government, identified by bracketed numbers for convenience in citation, are listed alphabetically under Nigeria.

Abernethy, David B. The Political Dilemma of Popular Education, An African Case. Stanford, Calif.: Stanford Univ. Press, 1969.
Aboyade, Ojetunji. Foundations of an African Economy. New York: Frederick A. Praeger, 1966.
Adelman, Morris A. "The World Oil Outlook," in Marion Clawson, ed., Natural Resources and International Development. Baltimore: Johns Hopkins Press, 1964, pp. 27–125.
Aluko, S. A., and M. O. Ijere. "The Economics of Mineral Oil," *Nigerian Journal of Economic and Social Studies*, VII (July 1965), 209–20.
Asiodu, P. C. "Industrial Development." Ibadan, Nigeria: Conference on National Reconstruction and Development in Nigeria, March 1969.
Baldwin, Robert E. Economic Development and Export Growth: A Study of Northern Rhodesia, 1920–1960. Berkeley: Univ. of Calif. Press, 1966.
Bardhan, Pranab. "External Economies, Economic Development, and the Theory of Protection." *Oxford Economic Papers*, n.s., XVI (March 1964), 40–54.
Bowles, Samuel. Planning Educational Systems for Economic Growth. Cambridge: Harvard Univ. Press, 1969.
Carter, Nicholas G. "An Input-Output Analysis of the Nigerian Economy, 1959–1960." Working Paper, School of Industrial Management, Massachusetts Institute of Technology, August 1963. (Mimeographed.) Summarized as an appendix in "An Input-Output Analysis of the Nigerian Economy, 1959–60," in Wolfgang F. Stolper, Planning Without Facts: Lessons in Resource Allocation from Nigeria's Development. Cambridge: Harvard Univ. Press, 1966, pp. 323–37.

Caves, Richard E. " 'Vent for Surplus' Models of Trade and Growth," in Robert E. Baldwin, Trade, Growth, and the Balance of Payments: Essays in Honor of Gottfried Haberler. Chicago: Rand McNally, 1965, pp. 95–115.

Chenery, Hollis B., and Paul G. Clark. Interindustry Economics. New York: John Wiley, 1959.

Chenery, Hollis B., and Alan M. Strout. Foreign Assistance and Economic Development. Discussion Paper No. 7. Washington, D.C.: U.S. Agency for International Development, June 1965.

Clark, Peter B. "The Choice of Optimal Import Substitution Patterns for Nigeria." Unpub. Ph.D. dissertation, Dept. of Economics, Massachusetts Institute of Technology, Cambridge, 1967.

Cownie, John B. "Nigerian National Income Accounts: Historical Summary and Projections to 1985." Consortium for the Study of Nigerian Rural Development, Michigan State Univ., East Lansing, August 1968.

Diallo, Mourtada. "Energy Resources and Utilisation." Ibadan, Nigeria: Conference on National Reconstruction and Development in Nigeria, March 1969.

Dickie, R. K. "Development of Crude Oil Production in Nigeria, and the Federal Government's Control Measures." Paper presented to the Institute of Petroleum, London, January 1966. (Mimeographed.)

Diejomaoh, Victor P. "Financing Development Expenditures: Nigerian Experience since 1950." Unpub. Ph.D. dissertation, Dept. of Economics, Harvard Univ., Cambridge, 1968.

Dorfman, Robert, Paul A. Samuelson, and Robert M. Solow. Linear Programming and Economic Analysis. New York: McGraw-Hill, 1958.

"Economic Growth of Nigeria: Problems and Prospects." 10 vols. Washington, D.C.: International Bank for Reconstruction and Development and International Development Association, November 1965. (Mimeographed.)

Eicher, Carl K., and Carl Liedholm, eds. Growth and Development of the Nigerian Economy. East Lansing, Mich.: Michigan State Univ. Press, 1970.

Energy Policy: Problems and Objectives. Paris: Organisation for Economic Co-operation and Development, 1966.

Frank, Charles R., Jr. "Industrialization and Employment Generation in Nigeria," Nigerian Journal of Economic and Social Studies, IX (November 1967), 277–97.

Frankel, Paul H. Essentials of Petroleum. London: Chapman and Hall, 1946.

Frankl, E. J., and S. A. Cordry. "The Niger Delta Oil Province—Recent Developments Onshore and Offshore." Paper presented to the Seventh World Petroleum Congress, Mexico City, April 1967. (Mimeographed.)

Gabriel, K. Georg. "The Gains to the Local Economy from the Foreign-owned Primary Export Industry: The Case of Oil in Venezuela." Unpub. Doctor of Business Administration dissertation, Graduate School of Business Administration, Harvard Univ., 1967.

Gray, Stanley. "The Oil Industry in Nigeria." Paper presented to the Bilateral Discussions between the Nigerian and United Kingdom Chambers of Commerce, December 1964. (Mimeographed.)

Harris, John R. "Socioeconomic Determinants of Entrepreneurial Success in Nigerian Industry." Paper presented to the Washington Meetings of the Econometric Society, December 1967, Massachusetts Institute of Technology, Cambridge, December 1967. (Mimeographed.)

Hartshorn, J. E. Politics and World Oil Economics: An Account of the International Petroleum Industry in Its Political Environment. New York: Frederick A. Praeger, 1967.

Hay, Alan M., and Robert H. T. Smith. "Preliminary Estimates of Nigeria's Interregional Trade and Associated Money Flows," *Nigerian Journal of Economic and Social Studies*, VIII (March 1966).

Helleiner, Gerald K. Peasant Agriculture, Government, and Economic Growth in Nigeria. Homewood, Ill.: Richard D. Irwin, 1966.

———. "Typology in Development Theory: The Land Surplus Economy (Nigeria)." *Food Research Institute Studies*, VI, No. 2 (1966), 181–94.

Hinrichs, Harley H. "Tax Strategies for Financing Development: The Nigerian Case." U.S. Agency for International Development, Washington, D.C., 1966. (Mimeographed.)

Hirschman, Albert O. Development Projects Observed. Washington, D.C.: Brookings Institution, 1967.

———. The Strategy of Economic Development. New Haven: Yale Univ. Press, 1958.

Johnson, Glenn L., et al. Strategies and Recommendations for Nigerian Rural Development, 1969/1985. East Lansing, Mich., Consortium for the Study of Nigerian Rural Development, 1969.

Johnston, Bruce F. "Agriculture and Structural Transformation in Developing Countries: A Survey of Research," *Journal of Economic Literature*, forthcoming, 1970.

———. "Agriculture's Role in Nigerian Development Strategy." U.S. Agency for International Development, Washington, D.C., 1966. (Mimeographed.)

Jones, William O. "Environment, Technical Knowledge and Economic Development in Tropical Africa." *Food Research Institute Studies*, V, No. 2 (1965), 101–16.

———. "Marketing of Staple Food Crops in Tropical Africa: Overall Analysis and Report." Food Research Institute, Stanford Univ., May 1969. (Mimeographed.)

Kilby, Peter. The Development of Small Industry in Eastern Nigeria. Washington, D.C.: U.S. Agency for International Development, 1962.
———. Industrialization in an Open Economy: Nigeria, 1945–1966. Cambridge, Eng.: Cambridge Univ. Press, 1969.
Lewis, W. Arthur. Reflections on Nigeria's Economic Growth. Paris: Development Centre of the Organisation for Economic Co-operation and Development, 1967.
McKinnon, Ronald I. "Foreign Exchange Constraints in Economic Development and Efficient Aid Allocations." *Economic Journal*, LXXIV (June 1964), 388–409.
Manne, Alan S., and Harry M. Markowitz, eds. Studies in Process Analysis, Economy-wide Production Capabilities. Cowles Foundation for Research in Economics at Yale Univ., Monograph No. 18. New York: John Wiley, 1963.

Nigeria, government publications. Printed in Lagos by the Government Printer unless otherwise noted.

[1] Central Bank of Nigeria. Annual Report and Statement of Accounts for the Year Ended 31st December 1965. 1966.
[2] ———. Annual Report and Statement of Accounts for the Year Ended 31st December 1966. 1967.
[3] Federal Ministry of Information. Annual Report of the Petroleum Division of the Federal Ministry of Mines and Power, 1962–63. 1964.
[4] ———. Annual Report of the Petroleum Division of the Federal Ministry of Mines and Power, 1963–64. 1965.
[5] ———. Annual Report of the Petroleum Division of the Federal Ministry of Mines and Power, 1964–65. 1966.
[6] ———. Annual Report of the Petroleum Division of the Federal Ministry of Mines and Power, 1965–66. 1967.
[7] ———. Annual Report of the Petroleum Division of the Federal Ministry of Mines and Power, 1966–67. 1968.
[8] ———. Monthly Petroleum Information: Ministry of Mines and Power. January 1968.
[9] Federal Ministry of Labour. Report on Employment and Earnings Enquiry, December 1962. 1964.
[10] Federal Office of Statistics. Annual Abstract of Statistics, 1965. 1967.
[11] ———. Annual Abstract of Statistics, 1966. 1968.
[12] ———. Annual Abstract of Statistics, 1967. 1969.
[13] ———. Digest of Statistics, XVI (April 1967).
[14] ———. Digest of Statistics, XVII (October 1968).
[15] ———. Economic Indicators, December 1967.

[16] ———. Economic Indicators, May 1968.
[17] ———. Review of External Trade, 1968.
[18] Federation of Nigeria. National Development Plan, 1962–68. Federal Ministry of Economic Development, 1963.
[19] ———. National Development Plan, Progress Report, 1964. Apapa, Nigeria: Federal Ministry of Economic Development, Nigerian National Press, 1965.
[20] Rivers State. Office of the Governor. Information Unit. The Oil Rich Rivers State, 1967.
"Nigeria: Growing Source of World Oil," *Institute of Petroleum Review*, July 1963, pp. 233–41.
"Nigeria Jumps Tax Ante on Oil," *Oil and Gas Journal*, LXV (January 27, 1967), 58–59.
Nigerian Human Resource Development and Utilization. New York: Education and World Affairs, 1967.
"Nigeria's Oil Revolution Begins," *West Africa*, February 27, 1965, pp. 233–35.
O'Connell, James. "Political Integration: The Nigerian Case," in Arthur Hazlewood, ed., African Integration and Disintegration. London: Oxford Univ. Press, 1967, pp. 129–84.
Oluwasanmi, H. A. Agriculture and Nigerian Economic Development. Ibadan: Oxford Univ. Press, 1966.
Pearson, Scott R. "The Impacts of Petroleum on the Nigerian Economy." Unpub. Ph.D. dissertation, Dept. of Economics, Harvard Univ., Cambridge, 1968.
———. "Nigerian Petroleum: Implications for Medium-Term Planning," in Carl K. Eicher and Carl Liedholm, eds., Growth and Development of the Nigerian Economy. East Lansing: Michigan State Univ. Press, 1970, pp. 352–75.
———. "The Political Economics of Nigerian Short-Term Borrowing," in *Public Policy*, XV, edited by John D. Montgomery and Arthur Smithies. Cambridge: Harvard Univ. Press, 1966, pp. 337–60.
Penrose, Edith T. The Large International Firm in Developing Countries: The International Petroleum Industry. London: George Allen and Unwin Ltd., 1968.
Reynolds, Clark W. "Development Problems of an Export Economy: The Case of Chile and Copper," in Markos Mamalakis and Clark W. Reynolds, Essays on the Chilean Economy. Homewood, Ill.: Richard D. Irwin, 1965, pp. 203–398.
Robinson, M. S. "Nigerian Oil: Prospects and Perspectives," *Nigerian Journal of Economic and Social Studies*, VI (July 1964), 219–29.
Roemer, Michael. "The Dynamic Role of Exports in Economic Develop-

ment: The Fishmeal Industry in Peru, 1956–1966." Unpub. Ph.D. dissertation, Dept. of Economics, Massachusetts Institute of Technology, 1967.

Schatzl, L. H. Petroleum in Nigeria. Ibadan, Nigeria: Oxford Univ. Press, published for the Nigerian Institute of Social and Economic Research, Univ. of Ibadan, 1969.

Schwarz, Walter. "Foreign Powers and the Nigerian War," *Africa Report*, XV (February 1970), 12–14.

———. Nigeria. New York: Frederick A. Praeger, 1968.

The Shell-BP Story. Port Harcourt, Nigeria: The Shell-BP Petroleum Development Co. of Nigeria Ltd., 1965.

Sklar, Richard L. Nigerian Political Parties: Power in an Emergent African Nation. Princeton, N.J.: Princeton Univ. Press, 1963.

———. "Nigerian Politics in Perspective," *Government and Opposition*, II, No. 2 (July/October 1967), 524–39.

Sokolski, Alan. The Establishment of Manufacturing in Nigeria. New York: Frederick A. Praeger, 1965.

Stolper, Wolfgang F. "Economic Growth and Political Instability in Nigeria: On Growing Together Again," in Carl K. Eicher and Carl Liedholm, eds., Growth and Development of the Nigerian Economy. East Lansing: Michigan State Univ. Press, 1970, pp. 328–51.

———. Planning Without Facts: Lessons in Resource Allocation from Nigeria's Development. Cambridge: Harvard Univ. Press, 1966.

———. "Prospects for the Nigerian Economy—Principles and Procedures Adopted in Projecting National Accounts." Nigerian National Press, Ltd., Apapa, Nigeria, 1962. (Mimeographed.)

Tanzer, Michael. The Political Economy of International Oil and the Underdeveloped Countries. New York: Beacon Press, 1969.

Vernon, Raymond. "Foreign-owned Enterprise in the Developing Countries," in *Public Policy*, XV, edited by John D. Montgomery and Arthur Smithies. Cambridge: Harvard Univ. Press, 1966, pp. 361–80.

Watkins, Melville H. "A Staple Theory of Economic Growth," *Canadian Journal of Economics and Political Science*, XXIX (May 1963), 141–58.

Index

Index

Afam, 96, 140
Afam Umuosi, 140
Africa, 19, 25
Agbada, 140
Agip, 14, 94, 162
Agricultural sector, 165–66
Agriculture: contribution to GDP, 32–33; exports, 32–33, 36–37, 38; development programs in, 35; imports, 81; figures for, 99, 122, 127, 131, 170ff, 173n, 195–200
Ahia, 140
Alesa-Eleme, 92
Algeria, 20, 134
Ancillary firms: analysis of role in economy, 68–69, 74, 82, 87, 97, 102; ownership of, 88–89; receipts of, 88, 90; salaries of, 99; survey of, 189–94

Balance-of-payments impacts, 75–79; international financial flows, 75f, 112–13; local currency expenditures, 75f, 111–12; figures for, 83, 174–75; estimation of impact of, 111–13; effect on economy, 113–16; Biafran, 148
Bauxite, Jamaican, 64
Belgium, 19
Benue-Plateau State, 16ff
Biafran economy, 145–52
Biafran secession, 2, 137–39, 156. See also Civil war
Binding factors: in calculating gain from foreign investment, 41, 155; identification of, 43–45; in input-output model, 50–54; of Nigerian economy, 70, 122–28, 130, 132

Bomu, 96, 140
Bonny, 179
BP, see British Petroleum
Brazil, 18f
British colonial government, 13
British government, 217n
British Petroleum, 15, 17, 92, 94

California Asiatic Oil Co., 17
Canada, 18f
Capital allowances: exhaustion of, 163, 184–85, 186; use of, 178–79, 182–83; generation of, 180–81; mentioned, 23f, 28–29, 71, 103, 114, 187
Capital flows: in petroleum industry's value added, 57, 117ff; in balance-of-payments impact, 112–13, 174–75; mentioned, 42, 60. See also Investment, foreign
Capital investment of petroleum industry, 74
Carter, Nicholas G., 125f
Chemical industries, 134–35
Chenery, Hollis B., 45
Chilean copper, 63–64
China, 6
Civil war: economic effects of, 55, 66, 95, 105ff, 117, 126; political effects of, 92, 156; mentioned, 14n, 15n, 35. See also Biafran secession
Cocoa, 31, 33f, 37
Commissioner of Mines and Power, 162, 164
Compagnie Française des Pétroles (Total), 94
Companies' Decree, 27–28, 161
Concessions, 13–17 *passim*, 30, 159, 176–77. See also Oil Exploration Li-

censes; Oil Mining Leases; Oil Prospecting Licenses
Conch Methane Services, 134
Continental Oil, 134
Contractors, 189–94. *See also* Ancillary firms
Copper, 63–64
Cotton, 31, 33–37 *passim*
Covenants, 22, 24–25
Cretaceous areas, 15
Crude petroleum, *see* Petroleum
Curacao, 19
Cumberland Corp., 17
Customs duties, 21–22, 111, 143f, 179

Decrees, 17n, 24, 143n, 180
Delta Oil (Nigeria) Ltd., 17n
Denmark, 19
Depletion allowances, 8
Depreciation, 63, 205–7. *See also* Capital allowances
Discounts, 178
Distributable Pool Account Decree of 1970, 143n
Drilling, 60, 62, 82, 89f, 178, 190–91

East Central State, 15ff, 144f, 139ff
Eastern Europe, 6
Eastern Nigeria Development Corporation, 96
Eastern Region, 139–45 *passim*, 149
Eastern States, 106–7
Ebubu, 140
Economic rents, 154
Economies of scale, 46ff, 86ff
Economy, *see* Nigerian economy
Education, 36, 49, 100f, 166
Electricity Corporation of Nigeria, 95f
Employment, 98–99, 157; of expatriates, 21, 84, 98–99, 100, 193–94; as measure of economic welfare, 39–40; levels of, 84; of Nigerians, 84, 98–99; data for Nigeria, 99; of Nigerians by ancillary firms, 193; mentioned, 38
Energy, 5f, 95–97
E.N.I. (Ente Nazionale Idracarbum), 16
Enterprise de Recherches et d'Activités Pétrolières (ERAP), 16
Eriemu, 140
Esso, 17n, 18, 94n

Expatriates, 21, 84, 98–99, 100, 193–94
Exploring and producing companies, 16–17. *See also* Petroleum industry
Exports, agricultural, 37
Exports, petroleum: figures for, 6, 19, 55–56, 94, 170ff, 173n, 174f; description of, 31, 34; earnings, 57, 113–19 *passim*; in balance-of-payments analysis, 75f, 112–13; mentioned, 5, 10
Externalities, 46, 48, 86f, 100

Factor constraint, *see* Binding factors
Factor contributions to other sectors, 70–85
Factor immobility, 48
Factor income paid abroad, 57, 75ff, 113, 117ff
Factor opportunity costs, *see* Opportunity costs
Factor payments, 49, 55, 59–66, 68, 119
Factor price distortions, 157
Factor shares, 59–66
Factors of production: economic effects of, 40–41, 47, 50, 70; creation of, 42–45; petroleum's contribution of scarce, 70–71, 73, 83, 85, 155, 157; availabilities, 126, 130; uses, 127, 131. *See also* Binding factors; foreign exchange; Investment resources; Labor
Federal Minister of Finance, 137
Federal Ministry of Finance, 26
Federal Ministry of Labour, 99
Federal Nigeria, *see* Nigerian Federation
Federal Office of Statistics, 65, 211 n5, 214 n1 (chap. 5), 216 n5
Federal Prime Minister, 137
Fertilizers, 135, 136n, 165, 217 n11
Field storage value, 177–78, 188
Fishing, 32, 170–73
Foreign exchange: as binding factor, 44–45, 66–67, 71, 128, 132, 155, 157; availability and use of, 70–71, 77–79, 94, 113, 114–15; contributed to other sectors, 71, 74–75; future importance to economy, 116–20, 124–25; Biafran, 148–50, 151; input-output data for, 201–4; mentioned, 8, 41f, 111, 158

Forestry, 32, 170–73
France, 19
French government, 14

Glass Factory, 96
Great Basins Oil Company of Los Angeles, 17n, 18
Great Depression, 31
Gross Domestic Product (GDP), 32, 56–58, 100, 170ff, 173n
Gross National Product (GNP), 56–58, 170ff, 173n
Groundnuts, 31, 33f, 36f
Gulf Oil Co. (Nigeria) Ltd., 14n, 16, 26, 56, 107
Gulf Oil Corp., 16, 18

Harbor dues, 179; costs of, 27, 60ff, 63, 82, 117ff; role in economy, 57, 67f, 107
Hausas, 139
Head offices, 12–13, 30, 104–5, 153
Host country governments, 7–9, 9–13, 159. *See also* Less developed countries

Ibos, 138, 139–42
Ihandiagu, 15
Immigration Act of 1963, 84
Imo River, 96, 140
Imports: duties on, 21–22; opportunity costs of, 41; rationed, 45, 67; economic effects of, 57, 75, 87, 100, 112–13, 117ff, 174f; industry's expenditures on, 60ff, 63f, 81; indirect, 79–83; of services, 76f, replaced by refinery production, 94, nonpetroleum, 115, figures for total Nigerian, 170ff, 173n; of ancillary firms, 193
Income tax, 9, 193
Income Tax (Amendment) Decree No. 65, 176
Indonesia, 20
Industries, local, 49, 96
Infrastructure, 35, 49, 100f
Input-output analysis: to calculate binding factors and shadow prices, 45, 50–54; of economy's capacity for linkages, 48; theoretical model of, 50–54; of indirect imports, 79–80; application of theory of, 104, 120–32; data for economy, 195–204

Interest, 63
Intersectoral relationships: resulting in linkage effects, 45, 86, 155; benefits of, 53, 87; difficulty of measuring, 86; between petroleum and rest of economy, 122; mentioned, 3, 46, 49. *See also* Linkage effects
Investment, foreign: benefits and costs of, 1, 2–3; initial attraction of, 5, 153; pessimism about, 11; other effects on economy, 39–42, 42–54, 55, 57, 75f, 86; defined, 39n; net of opportunity costs, 40–42, 55; host country's attitudes toward, 159; mentioned, 13, 53. *See also* Factor contributions to other sectors; Factors of production; Linkage effects
Investment by local government, 159–62
Investment resources: as factor constraint, 44–45, 130, measured by input-output analysis, 53, 111, 201–4; stemming from petroleum industry, 70, 71–74, 102, 109, 157f; in projections for Nigerian economy, 123–24, 125ff, 130f
Iran, 10
Iraq, 10
Isimiri, 140
Italian government, 14
Ivory Coast, 19

Jamaican bauxite, 64
Japan Consulting Institute, 94

Kainji Dam, 134f
Kokori, 140
Kuwait, 10
Kwara State, 15f

Labor, skilled: opportunity costs of, 42, 67; training of, 42f; as factor constraint, 44–45, 128, 132; industry's requirements of, 65–66, 84, 201–4; contributed to economy, 71, 83–85; salaries of, 100; in projections for economy, 124f, 126f, 130f; effect of education policies on, 166; provisions of Petroleum Decree 1969 for, 187–88; mentioned, 33, 35, 60, 107, 155, 157
Labor, unskilled, 42, 44, 84

Lagos, 93
Lagos Chamber of Commerce, 26
Lagos Federal Territory, 214 n1 (chap. 5)
Lagos State, 15ff
Laws pertaining to petroleum, 176. *See also* Decrees
Less developed countries, 1, 13, 39–54. *See also* Host country governments
Lever Brothers, 96
Lewis, W. Arthur, 126
Libya, 20, 25f, 134
Linkage effects, 45–54, 70, 86–103, 155, 157; backward, 46, 53, 87–92, 132–33, 155; final demand, 46–54 *passim*, 86, 97–100, 136, 155, 157; fiscal, 46–50 *passim*, 86, 101–3, 156ff; forward, 46–47, 48, 155; immediate, 46–49; investment, 46–47, 86, 87–97; technological, 46–49 *passim*, 86, 100–101, 136, 155, 157; future, 53–54, 132–36
Local currency expenditures, 60f, 79, 80–83, 107–9, 111–12. *See also* Harbor dues; Payments to government
Local industries, 49, 96
Local sales, 55–56, 57, 76f, 106, 117ff

Management, 13, 42f, 71, 84, 100f
Manufacturing, 33–35, 36; figures for, 32, 170ff, 173n; comparison with petroleum industry, 63, 65; salaries for, 99; projections for, 122, 127, 131; input-output data for, 195–200; mentioned, 38, 81, 216 n5 (chap. 7)
Marketing, 6, 8
Markets, 11, 13f, 18f, 45, 48–49, 160
Middle East, 18, 20, 25, 217 n3
Mid-Western Region, 139, 143
Mid-Western State, 16f, 93, 139ff, 144–47 *passim*
Mineral Oil Act of 1958, 176
Mobil Oil Co., 16, 94
Mobil Producing Nigeria Ltd., 14n, 15f, 18
Morocco, 19
Most-favored-company clause, 25, 27
Most-favored-nation clause, 24–25

National income: as welfare criterion, 39, 165; growth rate of, 44; and petroleum, 154, 157; figures for, 166, 170–73

Natural gas: local sales of, 55, 95–97, 112; costs of, 87, 178; use of, 92, 134–36, 165; new provisions for, 164, 187f
Netherlands, 19
Niger Delta, 15
Nigeria (less Biafra), 151–52
Nigeria Agip Oil Company Ltd., 14, 16, 18
Nigerian Bitumen Corporation, 15
Nigerian Breweries, 96
Nigerian currency, 75f, 89
Nigerian economic development, 1, 3, 31–38. *See also* Companies' Decree
Nigerian economy: resilience of, 2; sectoral relationships in, 3, 10–11, 31f, 165–66; growth of, 38, 129–30, 158; significant measure of, 113–14; binding factors of, 122–28, 130, 132, 156; input-output data for, 195–204. *See also* Agriculture; Balance-of-payments impacts; Fishing; Forestry; Investment resources; Nigerian government; Petroleum industry; Utilities
Nigerian Federation, 138, 145, 149, 152
Nigerian government: policy-making processes of, 3, 35–36, 72–73, 111, 153–66; and petroleum industry, 12–13, 22–29, 105, 187–88; imposition of OPEC terms by, 24–26; revenues of, 73, 110; and refineries, 92, 93–94; participation in oil production of, 158; mentioned, 2, 14, 33. *See also* Host country governments; Payments to government; Saving, public
Nigerian Petroleum Refining Company, 92, 94, 96
Nigerian political system, 1f
Nigerian Ports Authority, 81, 83, 107f, 179
Nigerian subsidiaries, 12–13, 14, 16–17, 18, 30, 94. *See also* Companies' Decree; Gulf Oil Co. (Nigeria) Ltd.; Mobil Producing Nigeria Ltd.; Safrap; Shell-BP
Nkali, 140
North America, 18f
Northerners, 138
Northern Region, 143f, 214n
Norway, 19

Obagi, 140

Index

Obigbo North, 140
Offshore Terminal Dues Decree, 109, 179
Oil, see Petroleum
Oil Exploration Licenses (OEL's), 15–18 passim, 176, 188
Oil Mining Leases (OML's), 15–17 passim, 30n, 164, 177f, 187–88
Oil Pipelines Acts, 176
Oil Prospecting Licenses (OPL's), 15–18 passim, 164, 176–78 passim, 187f
Ojukwu, Colonel, 217 n3
Okan, 140
Oligopoly, 8
Oloibiri, 15, 140
Olomoro, 140
OPEC, 210 n12
OPEC (Organization of Petroleum Exporting Countries) terms: imposed, 24–26; illustration of, 28–29; effects of, 161–64 passim, 177ff, 184–87
Opportunity costs: social, 40, 55, 58f; measurement of, 41–42, 50, 54, 86; of salaries, 42; factor, 48, 66–69, 86; in investment linkages, 87f; in refinery operations, 94–95; of natural gas, 96; mentioned, 3, 46, 53, 91
Organization for Economic Co-operation and Development (OECD), 5f, 6n
Oweh, 140
Ownership, by host country, 159–60

Palm produce, 31, 33f, 36f
Pay-as-you-earn withholding tax, 102, 193
Payments to government, 109–11, 176–88; in selected countries, 10; and other financial arrangements, 22–29, 105; industry expenditures, 28–29, 60–64 passim; benefits of, 49–50, 70, 101–2; and petroleum industry value added, 57; opportunity costs of, 67ff; and total revenues, 71–72, 110–11; total for 1965, 102; origin and allocation of, 138–39, 142–45; and Biafran revenues, 147; mentioned, 66, 75, 80–83 passim, 87, 103, 117ff, 155, 158. See also Rentals; Royalties; Taxes
Persian Gulf crude, 18n
Petrochemical industries, 134–35
Petrochemicals, 165, 217 n11
Petroleum: discovery of, 1; political implications of, 2, 156, 159, 176; demand for, 5, 11, 30; location of, 18, 139–42; quality of, 18–20; exploration for, 57; Venezuelan, 63–64; and civil war, 138–39; ethnic origin of, 138–42; historical impact of, 153–56; future impact of, 156–58. See also Political considerations; Projections, for petroleum
Petroleum corporations, international, 1, 3, 4–11, 83–84. See also Companies' Decree; Head offices; Nigerian subsidiaries
Petroleum Decree 1969, 14, 30n, 162, 164, 176, 187–88
Petroleum industry: and economy, 3, 9–13, 32, 55–69; vertical integration in, 6ff; cost-price structure of, 7–9, 22–23; relationship with government, 12–13, 20–22; factors influencing operations of, 18–22; financial arrangements with government, 22–29, 105; total receipts of, 59; Leontief production functions for, 60, 63; coefficients, 62–65, 67–69; comparison with foreign-owned industries, 63–64; factors of production of, 104–21, 121–32; balance-of-payments impact of, 111–14, 116; and foreign exchange availability, 114–16; value added of, 116–20, 121; and Biafran foreign exchange availability, 148–52; input-output data for, 195–204; mentioned, 74, 98–99, 108. See also Ancillary firms; Balance-of-payments impacts; Drilling; Employment; Exports, petroleum; Imports; Nigerian subsidiaries
Petroleum production: world totals, 6; phases of, 7; and payments to government, 10; levels of, 30, 55–56, 57, 106; divided between Biafra and Mid-Western Nigeria, 147; mentioned, 1, 3, 5. See also Payments to government
Petroleum products, 162
Petroleum Profits Tax (Amendment) Decree, 176
Petroleum Profits Tax Ordinance of 1959, 23, 176, 182
Phillips Oil Co. (Nigeria) Ltd., 14, 16

Phillips Petroleum Co., 16, 18
Pipeline application leases, 20
Policy issues, 4, 159–66
Political considerations, 4, 40, 137, 142–45, 160, 166
Port charges, see Harbor dues
Port Harcourt, 15, 85, 92, 95f
Posted price, 25, 27; discounts, 24n, 27; in calculating petroleum value added, 57; and tax revenues, 163; in OPEC terms, 184f; mentioned, 9, 29, 123
Premiums, 72
Producing countries, see Host country governments
Profit, chargeable, 24, 28–29, 185, 278–79
Profits: expatriated, 98, 193; mentioned, 8, 21, 28–29, 63, 87, 155, 205–7
Profits tax, 109, 143f; legislation, 21–25 passim, 176–79 passim; in OPEC terms, 185–86
Projections: for petroleum in Nigeria, 3–4, 104–36; of hypothetical Biafran economy, 145–52
Protection measures taken by importers, 8f
Public saving, see Saving, public
Public utilities, see Utilities
Puerto Rico, 19

Realized price, 23, 27, 28–29, 57, 129, 163, 177
Refinery, 55, 93–95, 112, 133–34, 162
Refining, 6, 160
Remuekpe, 140
Rentals, 23, 72, 143f, 177, 179
Rents, 154n, 188
Requirements analysis, see Input-output analysis
Resources, 40–48 passim, 86f
Revenues, see Payments to government; Taxes
Reynolds, Clark W., 64n, 214 n1
Rivers State, 15ff, 92, 139–44 passim
Royal Dutch Shell Petroleum Co., 15, 17, 92
Royalties: expensing, 25, 184–87; annual table of, 109; in OPEC terms, 163, 177–78; in tax calculation, 178f, 184; mentioned, 23, 72, 143f, 188

Rubber, 31–37 passim

Safrap (Nigeria) Ltd., 14n, 15n, 16, 18, 26–56
Salaries: opportunity costs of, 42, 67f; in calculating petroleum industry's value added, 57; industry expenditures, 60–64 passim, 98–99, 125, 136; expatriate, 193f; paid to Nigerians by ancillary firms, 193; mentioned, 65–66, 75, 81f, 87, 97, 107, 117ff, 155, 205–7
Saudi Arabia, 10
Saving, public: benefited by petroleum industry, 43, 70; growth of, 73–74, 111; mentioned, 102, 109, 155, 158, 170ff, 173n
Schmidt, Wilson E., 189, 192
Secession, see Biafran secession
Sensitivity analysis, 50, 53, 130–32
Services, input-output data for, 195–200
Services sector of economy, 122, 127, 131
Servicing and supplying firms, see Ancillary firms
Shadow prices, 43–45
Shell, 94, 134. See also Royal Dutch Shell; Shell-BP
Shell-BP Petroleum Company of Nigeria Ltd.: reducing costs, 22n–23n; leader among Nigerian companies, 26; and NPRC refinery crude, 94; legalities of Nigerian concessions, 210 n5; mentioned, 14–18 passim, 56, 85, 92n, 95, 106, 215 n5, 217 n3
Shell/D'Arcy Exploration Parties, 15
Sinclair International Oil Co., 17
Six-Day War, 217 n3
Société Africaine d'Exploration Pétrolière (SAFREX), 16
Société de Gestion des Participations de la Régie Autonome des Pétroles (SOGERAP), 16
South America, 18f
South Eastern State, 16f, 139, 144f
Spain, 19
Stamp, L. Dudley, 13n
Standard of New Jersey, 14, 17n
Stolper, Wolfgang F., 156
Strout, Alan M., 45
Suez Canal, 18

Index

Sunray DX Oil Co., 17
Suppliers, 189–94. *See also* Ancillary firms
Sweden, 19

Taxes: policy, 7–9, 162–63, 176–79; dominating cost factors, 22–23; effect of OPEC terms, 26, 28–29, 185–86; affecting agriculture, 36–37; PAYE, 102; ancillary firms', 102–3; comparison with competitive countries, 159–61; figures for, 170ff, 173n; mentioned, 63, 65, 193. *See also* Capital allowances; Payments to government; Profit, chargeable; Realized price; Rentals
Technology, 13, 100f
Tenneco Oil Co. of Nigeria Ltd., 17f
Tennessee Gas Transmission Co., 17
Texaco, 14n, 94
Texas Overseas Petroleum Co., 17f
Textile Mills, 96
Training: programs in, 84–85, 165; projection for, 136; future provisions for, 164; mentioned, 42f, 49, 100, 155, 157
Trans-Amadi Industrial Estate, 95f
Transfer prices, 7–9, 123, 154, 174–75
Transportation, 6, 160, 190–91

Ughelli, 96, 140
Umuechem, 140

Underdeveloped countries, *see* Less developed countries
Unemployment, 38, 157
Union Oil Co. of California, 17
Union Oil Nigeria, 17
Union Stockyard and Transit Company of Chicago, 134
United Kingdom, 18f, 75n, 134
United Kingdom Gas Council, 134
United States, 9, 16–20 *passim*
Uruguay, 18f
U.S. Agency for International Development Mission to Nigeria, 189
U.S. crude, 8
USSR, 6, 9, 20
Utilities, 32, 35, 96, 170ff, 173n. *See also* Services sector of economy
Uzere East, 140
Uzere West, 140

Venezuela, 10, 18, 20
Venezuelan petroleum, 63–64

Wages, *see* Salaries
Western Europe, 18f
Western Germany, 19
Western Region, 143f, 214 n1 (chap. 5)
Western State, 16f
West Indies, 19
Wilson, Harold, 217 n3

Zambian copper, 63–64

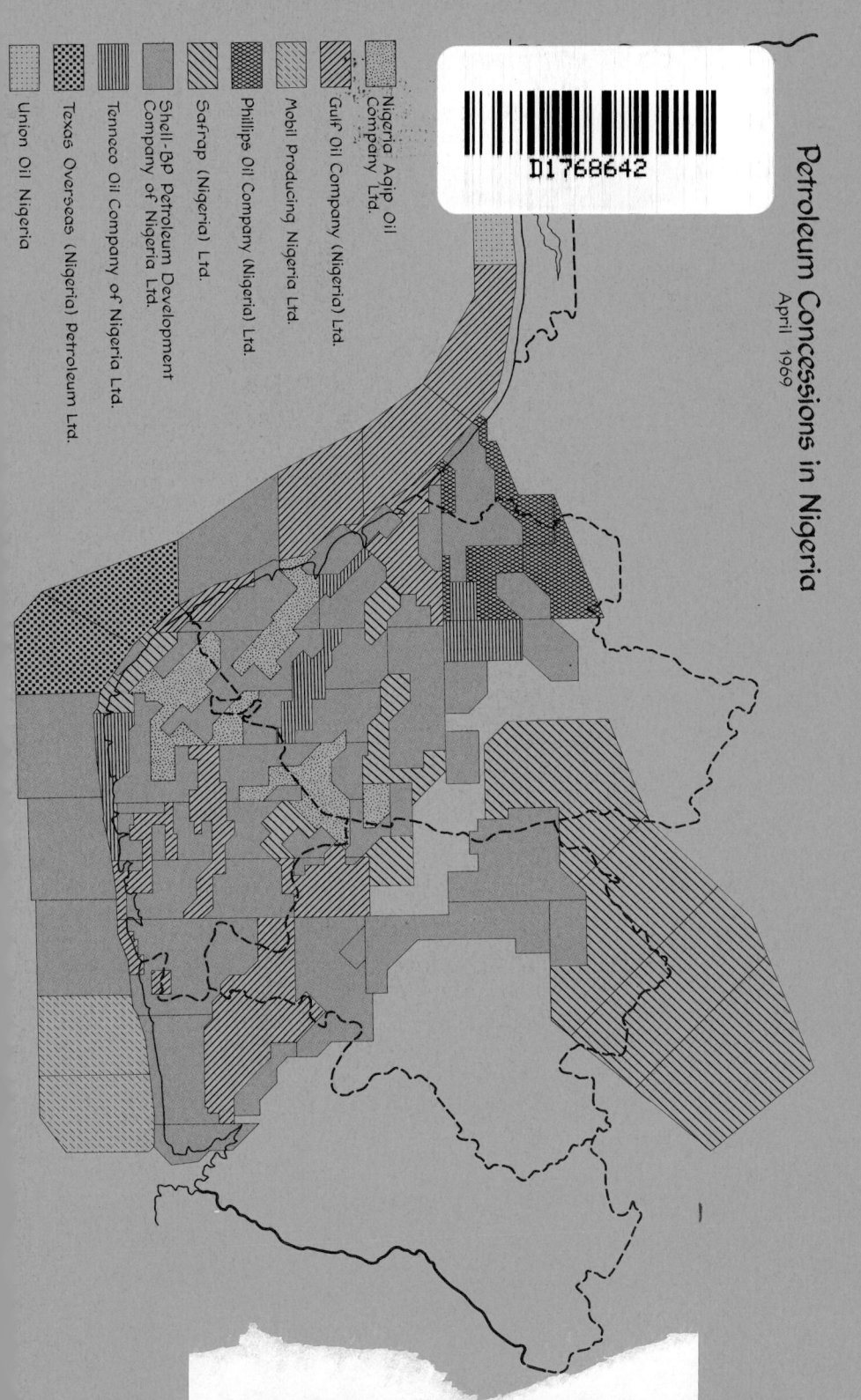